Praise for

Temple Grandin

"What do neurologists, cattle, and McDonald's have in common? They all owe a great deal to one woman, a renowned animal scientist born with autism, Temple Grandin. . . . an extraordinary source of inspiration for autistic children, their parents — and all people."

— *Time*, "The 2010 *Time* 100 Most Influential People in the World"

"We are lucky to have Temple Grandin." ***— New York Times***

Praise for

The Autistic Brain

"Grandin has reached a stunning level of sophistication about herself and the science of autism. Her observations will assist not only fellow autistics and families with affected members, but also researchers and physicians seeking to better understand the condition."

— Jerome Groopman, *New York Review of Books*

"Her visual circuitry extends well beyond where neurotypicals' circuitry stops. Grandin is wired for long-term visual memory. She is sure that one day, autism will be explained by neurobiology. Her new book, *The Autistic Brain,* outlines that quest." ***— Los Angeles Times***

"Grandin has helped us understand autism not just as a phenomenon, but as a different but coherent mode of existence that otherwise confounds us. . . . She excels at finding concrete examples that reveal the perceptual and social limitations of autistic and 'neurotypical' people alike." ***New York Times***

"Grandin's particular skill is her remarkable ability to make sense of autistics' experiences, enabling readers to see 'the world through an autistic person's jumble of neuron misfires,' and she offers hope that one day, autism will be considered not according to some diagnostic manual, but to the individual." — ***Publishers Weekly***

"Autism is a spectrum, and Temple is on one edge. Living on this edge has allowed her to be an extraordinary source of inspiration for autistic children, their parents—and all people." — ***Time***

"An iconic example of someone who puts her strengths, and even her limitations, to good use." — **KQED, San Francisco**

"Temple Grandin has yet again been of enormous service to the millions of autistic individuals worldwide, to anyone labeled with a disability, and to the rest of us curious about the brain and the intricacies of human experience."

— ***New York Journal of Books***

"Highly recommended for anyone who knows or works with people on the spectrum." — ***Library Journal,*** **starred review**

"*The Autistic Brain* is an engaging look at life within the spectrum. It's also an honest one." — ***HealthCare Book Reviews***

"A tremendous gift, not just to patients and their families, but also to teachers, mentors, friends, and everyone who is interested in understanding how our brains make us who we are . . . This is a book everyone should read."

— **Dr. Ginger Campbell, *Brain Science Podcast***

"An important and ultimately optimistic work." —***Booklist***

"An illuminating look at how neuroscience opens a window into the mind." —***Kirkus***

Praise for

Animals in Translation

"Neurology has Oliver Sachs, nature has Annie Dillard, and the lucky animal world has Grandin, a master intermediary between humans and our fellow beasts . . . At once hilarious, fascinating, and just plain weird, *Animals* is one of those rare books that elicits a 'wow' on almost every page. A." —***Entertainment Weekly***

"Grandin's focus in *Animals in Translation* is not on all the 'normal' things autistics and animals can't do but on the unexpected, extraordinary, invaluable things they can." —***O, the Oprah Magazine***

"Provocative, clear, and funny, *Animals in Translation* calls into question who and what is 'normal' . . . What Grandin lays out so clearly by using her own autism to understand animals has huge implications for the way we live with and without animals." —***Minneapolis Star Tribune***

"Full of heart, soul, and crackling intelligence." —***People***

"Inspiring . . . Crammed with facts and anecdotes about Temple Grandin's favorite subject: the senses, brains, emotions, and amazing talents of animals." —***New York Times Book Review***

Praise for

Animals Make Us Human

"Most animal lovers will come away from this book making some practical adjustment in the daily life of the animal closest to them. And yet, ultimately, this book is much more than a simple primer. Species by species, insight by insight, it builds into something kind of momentous: the realization or confirmation (depending on your perspective) that animals possess very complex minds and that science is just beginning to provide tiny glimpses of this uncharted territory."

—Boston Globe

"A practical, species-by-species guide to making animals happier."

—Time

"We are lucky to have Temple Grandin."

—New York Times

"For pet owners, Temple Grandin's perspective is invaluable."

—Entertainment Weekly

"Sure to become a classic in the field of human/animal relationships."

—Patricia McConnell, author of *For the Love of a Dog* and *The Other End of the Leash*

"An illuminating look at how neuroscience opens a window into the mind."

—Kirkus

The Autistic Brain

Also by Temple Grandin

Animals Make Us Human: Creating the Best Life for Animals (with Catherine Johnson)

Animals in Translation: Using the Mysteries of Autism to Decode Animal Behavior (with Catherine Johnson)

Emergence: Labeled Autistic (with Margaret M. Scariano)

Thinking in Pictures: And Other Reports from My Life with Autism

Genetics and the Behavior of Domestic Animals (editor)

Livestock Handling and Transport (editor)

Developing Talents (with Kate Duffy)

Humane Livestock Handling (with Mark Deesing)

The Way I See It: A Personal Look at Autism and Asperger's

Improving Animal Welfare: A Practical Approach (editor)

Different . . . Not Less: Inspiring Stories of Achievement and Successful Employment from Adults with Autism, Asperger's, and ADHD (editor)

Also by Temple Grandin and Sean Barron

Unwritten Rules of Social Relationships

Also by Richard Panek

The 4 Percent Universe: Dark Matter, Dark Energy, and the Race to Discover the Rest of Reality

The Invisible Century: Einstein, Freud, and the Search for Hidden Universes

Seeing and Believing: How the Telescope Opened Our Eyes and Minds to the Heavens

Waterloo Diamonds

The Autistic Brain

Helping Differents Kinds of Minds Succeed

TEMPLE GRANDIN

and RICHARD PANEK

MARINER BOOKS
HOUGHTON MIFFLIN HARCOURT
BOSTON • NEW YORK

First Mariner Books edition 2014

www.hmhco.com

Library of Congress Cataloging-in-Publication Data
Grandin, Temple.
The autistic brain : thinking across the spectrum /
Temple Grandin and Richard Panek.
pages cm
ISBN 978-0-547-63645-0 (hardback) ISBN 978-0-544-22773-6 (pbk.)
1. Autism spectrum disorders. 2. Autistic people — Mental health.
3. Autism — Research. 4. Psychology, Pathological. I. Panek, Richard. II. Title.
RC553.A88G725 2013
616.85'882 — dc23 2013000662

Printed in the United States of America
DOC 10 9 8 7 6 5 4 3 2 1

Contents

Prologue *vii*

The Autistic Brain

1. The Meanings of Autism 3
2. Lighting Up the Autistic Brain 21
3. Sequencing the Autistic Brain 50
4. Hiding and Seeking 69

Rethinking the Autistic Brain

5. Looking Past the Labels 101
6. Knowing Your Own Strengths 117
7. Rethinking in Pictures 134
8. From the Margins to the Mainstream 171

Appendix: The AQ Test *207*

Notes *217*

Acknowledgments *229*

Index *231*

Prologue

IN THIS BOOK I will be your guide on a tour of the autistic brain. I am in the unique position to speak about both my experiences with autism and the insights I have gained from undergoing numerous brain scans over the decades, always with the latest technology. In the late 1980s, shortly after MRI became available, I jumped at the opportunity to travel on my first "journey to the center of my brain." MRI machines were rarities in those days, and seeing the detailed anatomy of my brain was awesome. Since then, every time a new scanning method becomes available, I am the first in line to try it out. My many brain scans have provided possible explanations for my childhood speech delay, panic attacks, and facial-recognition difficulties.

Autism and other developmental disorders still have to be diagnosed with a clumsy system of behavioral profiling provided in a book called the *DSM*, which is short for *Diagnostic and Statistical Manual of Mental Disorders*. Unlike a diagnosis for strep throat, the diagnostic criteria for autism have changed with each new edition of the *DSM*. I warn parents, teachers, and therapists to avoid getting locked into the labels. They are not precise. I beg you: Do not allow a child or an adult to become defined by a *DSM* label.

The genetics of autism is an exceedingly complex quagmire. Many small variations in the genetic code that control brain development are involved. A genetic variation that is found in one autistic child will be absent in another autistic child. I will review the latest in genetics.

Researchers have done hundreds of studies on autistics' problems with social communication and facial recognition, but they have neglected sensory issues. Sensory oversensitivity is totally debilitating for some people and mild in others. Sensory problems may make it impossible for some individuals on the autism spectrum to participate in normal family activities, much less get jobs. This is why my top priorities for autism research are accurate diagnoses and improved treatments for sensory problems.

Autism, depression, and other disorders are on a continuum ranging from normal to abnormal. Too much of a trait causes severe disability, but a little bit can provide an advantage. If all genetic brain disorders were eliminated, people might be happier, but there would be a terrible price.

When I wrote *Thinking in Pictures,* in 1995, I mistakenly thought that everybody on the autism spectrum was a photorealistic visual thinker like me. When I started interviewing other people about how they recalled information, I realized I was wrong. I theorized that there were three types of specialized thinking, and I was ecstatic when I found several research studies that verified my thesis. Understanding what kind of thinker you are can help you respect your limitations and, just as important, take advantage of your strengths.

The landscape I was born into sixty-five years ago was a very different place from where we are now. We've gone from institutionalizing children with severe autism to trying to provide them with the most fulfilling lives possible — and, as you will read in chapter 8, finding meaningful work for those who are able. This book will show you every step of my journey.

— TG

Part I

THE AUTISTIC BRAIN

1 The Meanings of Autism

I WAS FORTUNATE to have been born in 1947. If I had been born ten years later, my life as a person with autism would have been a lot different. In 1947, the diagnosis of autism was only four years old. Almost nobody knew what it meant. When Mother noticed in me the symptoms that we would now label autistic — destructive behavior, inability to speak, a sensitivity to physical contact, a fixation on spinning objects, and so on — she did what made sense to her. She took me to a neurologist.

Bronson Crothers had served as the director of the neurology service at Boston Children's Hospital since its founding, in 1920. The first thing Dr. Crothers did in my case was administer an electroencephalogram, or EEG, to make sure I didn't have petit mal epilepsy. Then he tested my hearing to make sure I wasn't deaf. "Well, she certainly is an odd little girl," he told Mother. Then when I began to verbalize a little, Dr. Crothers modified his evaluation: "She's an odd little girl, but she'll learn how to talk." The diagnosis: brain damage.

He referred us to a speech therapist who ran a small school in the basement of her house. I suppose you could say the other kids there were brain damaged too; they suffered from Down syndrome and other disorders. Even though I was not deaf, I had difficulty hearing consonants, such as the *c* in *cup.* When grownups talked fast, I heard only the vowel sounds, so I thought they had their own special language. But by speaking slowly, the speech therapist helped me to hear the hard consonant sounds, and when I said *cup* with a *c*, she praised me — which is just what a behavioral therapist would do today.

At the same time, Mother hired a nanny who played constant turn-taking games with my sister and me. The nanny's approach was also similar to the one that behavioral therapists use today. She made sure that every game the three of us played was a turn-taking game. During meals, I was taught table manners, and I was not allowed to twirl my fork around over my head. The only time I could revert back to autism was for one hour after lunch. The rest of the day, I had to live in a nonrocking, nontwirling world.

Mother did heroic work. In fact, she discovered on her own the standard treatment that therapists use today. Therapists might disagree about the benefits of a particular aspect of this therapy versus a particular aspect of that therapy. But the core principle of every program — including the one that was used with me, Miss Reynolds's Basement Speech-Therapy School Plus Nanny — is to engage with the kid one-on-one for hours every day, twenty to forty hours per week.

The work Mother did, however, was based on the initial diagnosis of brain damage. Just a decade later, a doctor would probably have reached a completely different diagnosis. After examining me, the doctor would have told Mother, "It's a psychological problem — it's all in her mind." And then sent me to an institution.

While I've written extensively about autism, I've never really written about how the diagnosis itself is reached. Unlike meningitis or lung cancer or strep throat, autism can't be diagnosed in the laboratory — though researchers are trying to develop methods to do so, as we'll see later in this book. Instead, as with many psychiatric syndromes, such as depression and obsessive-compulsive disorder, autism is identified by observing and evaluating behaviors. Those observations and evaluations are subjective, and the behaviors vary from person to person. The diagnosis can be confusing, and it can be vague. It has changed over the years, and it continues to change.

The diagnosis of autism dates back to 1943, when Leo Kanner, a physician at Johns Hopkins University and a pioneer in child psychiatry, proposed it in a paper. A few years earlier, he had received a letter from a worried father named Oliver Triplett Jr., a lawyer in Forest, Mississippi. Over the course of thirty-three pages, Triplett described in detail the first five years of his son Donald's life. Donald, he wrote, didn't show signs of wanting to be with his mother, Mary. He could be "perfectly oblivious" to everyone else around him too. He had frequent tantrums, often didn't respond to his name, found spinning objects endlessly fascinating. Yet for all his developmental problems, Donald also exhibited unusual talents. He had memorized the Twenty-Third Psalm ("The Lord is my shepherd...") by the age of two. He could recite twenty-five questions and answers from the Presbyterian catechism verbatim. He loved saying the letters of the alphabet backward. He had perfect pitch.

Mary and Oliver brought their son from Mississippi to Baltimore to meet Kanner. Over the next few years, Kanner began to identify in other children traits similar to Donald's. Was there a pattern? he wondered. Were these children all suffering from the same syndrome? In 1943, Kanner published a paper, "Autistic Disturbances of Affective Contact," in the journal *Nervous Child.* The paper presented the case histories of eleven children who, Kanner felt, shared a set of symptoms — ones that we would today recognize as consistent with autism: the need for solitude; the need for sameness. To be alone in a world that never varied.

From the start, medical professionals didn't know what to do with autism. Was the source of these behaviors biological, or was it psychological? Were these behaviors what these children had brought into the world? Or were they what the world had instilled in them? Was autism a product of nature or nurture?

Kanner himself leaned toward the biological explanation of autism, at least at first. In that 1943 paper, he noted that autistic behaviors seemed to be present at an early age. In the final paragraph, he wrote, "We must, then, assume that these children have come into the world with innate inability to form the usual, biologically provided affective contact with people, just as other children come into the world with innate physical or intellectual handcaps [*sic*]."

One aspect of his observations, however, puzzled him. "It is not easy to evaluate the fact that all of our patients have come of highly intelligent parents. This much is certain, that there is a great deal of obsessiveness in the family background"—no doubt thinking of Oliver Triplett's thirty-three-page letter. "The very detailed diaries and reports and the frequent remembrance, after several years, that the children had learned to recite twenty-five questions and answers of the Presbyterian Catechism, to sing thirty-seven nursery songs, or to discriminate between eighteen symphonies, furnish a telling illustration of parental obsessiveness.

"One other fact stands out prominently," Kanner continued. "In the whole group, there are very few really warmhearted fathers and mothers. For the most part, the parents, grandparents, and collaterals are persons strongly preoccupied with abstractions of a scientific, literary, or artistic nature, and limited in genuine interest in people."

These observations of Kanner's are not as damning about parents as they might sound. At this early point in his study of autism, Kanner wasn't necessarily suggesting cause and effect. He wasn't arguing that when the parents behaved *this way*, they caused their children to behave *that way*. Instead, he was noting similarities between the parents and his patients. The parents and their child, after all, belonged to the same gene pool. The behaviors of both generations could be due to the same biological hiccup.

In a 1949 follow-up paper, however, Kanner shifted his attention from the biological to the psychological. The paper was ten and a half pages long; Kanner spent five and a half of those pages on the behavior of the parents. Eleven years later, in an interview in *Time,* he said that autistic children often were the offspring of parents "just happening to defrost enough to produce a child." And since Kanner was the first and foremost expert on the subject of autism, his attitude shaped how the medical profession thought about the subject for at least a quarter of a century.

Late in life, Kanner maintained that he "was misquoted often as having said that 'it is all the parents' fault.'" He also complained that critics overlooked his original preference for a biological explanation. And he himself was no fan of Sigmund Freud; in a book he published in 1941, he wrote, "If you want to go on worshipping the Great God Unconscious and His cocksure interpreters, there is nothing to keep you from it."

But Kanner was also a product of his time, and his most productive years coincided with the rise of psychoanalytic thought in the United States. When Kanner looked at the effects of autism, he might have originally told himself that they were possibly biological in nature, but he nonetheless wound up seeking a psychological cause. And when he speculated on what villains might have inflicted the psychic injury, he rounded up psychoanalysis's usual suspects: the parents (especially Mom).

Kanner's reasoning was probably complicated by the fact that the behavior of kids who are the product of poor parenting can look like the behavior of kids with autism. Autistic kids can seem rude when they're actually just oblivious to social cues. They might throw tantrums. They won't sit still, won't share their toys, won't stop interrupting adult conversations. If you've never studied the behaviors of children with autism, you could easily conclude that these kids' parents are the problem, not the kids themselves.

But where Kanner went horribly wrong was in his assumption that because poor parenting can lead to bad behavior, all bad behavior must therefore be the result of poor parenting. He assumed that a three-year-old's ability to name all the U.S. presidents and vice presidents couldn't *not* be caused by outside intervention. He assumed that a child's psychically isolated or physically destructive behavior couldn't *not* be caused by parents who were emotionally distant.

In fact, Kanner had cause and effect backward. The child wasn't behaving in a psychically isolated or physically destructive manner because the parents were emotionally distant. Instead, the parents were emotionally distant because the child was behaving in a psychically isolated or physically destructive manner. My mother is a case in point. She has written that when I wouldn't return her hugs, she thought, *If Temple doesn't want me, I'll keep my distance.* The problem, though, wasn't that I didn't want her. It was that the sensory overload of a hug shorted out my nervous system. (Of course, nobody back then understood about sensory oversensitivity. I'll talk about this topic in chapter 4.)

Kanner's backward logic found its greatest champion in Bruno Bettelheim, the influential director of the University of Chicago's Orthogenic School for disturbed children. In 1967 he published *The Empty Fortress: Infantile Autism and the Birth of the Self,* a book that popularized Kanner's notion of the refrigerator mother. Like Kanner, Bettelheim thought that autism was probably biological in nature. And like Kanner, his thinking on autism was nonetheless grounded in psychoanalytic principles. Bettelheim argued that an autistic child was not biologically *predetermined* to manifest the symptoms. Instead, the child was biologically *predisposed* toward those symptoms. The autism was latent—until poor parenting came along and breathed life into it.[1]

If Mother hadn't taken me to a neurologist, she might eventually have been vulnerable to the refrigerator-mother guilt trip. She was

1. In the decade following Bettelheim's death in 1990, his reputation unraveled. Evidence emerged that he had misrepresented his education, plagarized, conducted shoddy research, and lied about being a doctor, but even more damning were accusations of physical and mental abuse by former students at the Orthogenic School.

only nineteen when I was born, and I was her first child. Like many young first-time mothers who find themselves confronting a child's "bad" behavior, Mother initially assumed she must be doing something wrong. Dr. Crothers, however, relieved that anxiety. When I was in second or third grade, Mother did get the full Kanner treatment from a doctor who informed her that the cause of my behavior was a psychic injury and that until I could identify it, I was doomed to inhabit my own little world of isolation.

But the problem wasn't a psychic injury, and Mother knew it. The psychoanalytic approach to a disorder was to find the cause of a behavior and try to remove it. Mother assumed she couldn't do anything about the cause of my behavior, so her approach was to concentrate on dealing with the behavior itself. In this respect, Mother was ahead of her time. It would take child psychiatry decades to catch up with her.

People often ask me, "When did you really know you were autistic?" As if there were one defining moment in my life, a before-and-after revelation. But the conception of autism in the early 1950s didn't work that way. Like me, child psychiatry back then was still young. The words *autism* and *autistic* barely appeared in the American Psychiatric Association's initial attempt to standardize psychiatric diagnoses, in the first edition of the *Diagnostic and Statistical Manual of Mental Disorders* (*DSM*), published in 1952, when I was five. The few times those words did appear, they were used to describe symptoms of a separate diagnosis, schizophrenia. For instance, under the heading Schizophrenic Reaction, Childhood Type, there was a reference to "psychotic reactions in children, manifesting primarily autism"—without further explanation of what autism itself was.

Mother remembers one of the early doctors in my life making a passing reference to "autistic tendencies." But I myself didn't actually hear the word *autistic* applied to me until I was about twelve or thirteen; I remember thinking, *Oh, it's me that's different.* Even then, though, I still wouldn't have been able to tell you exactly what autistic behaviors were. I still wouldn't have been able to tell you why I had such trouble making friends.

As late in life as my early thirties, when I was pursuing my doctorate at the University of Illinois at Urbana-Champaign, I could still overlook the role that autism played in my life. One of the requirements was a statistics course, and I was hopeless. I asked if I could take the course with a tutor instead of in a classroom, and I was told that in order to get permission to do that, I would have to undergo a "psychoeducational assessment." On December 17 and 22, 1982, I met with a psychologist and took several standard tests. Today, when I dig that report out of a file and reread it, the scores practically scream out at me, *The person who took these tests is autistic.*

I performed at the second-grade level on a subtest that required me to identify a word that was spoken at the rate of one syllable per second. I also scored at the second-grade level on a subtest that required me to understand sentences where arbitrary symbols replaced regular nouns — for instance, a flag symbol meant "horse."

Well, yeah, I thought, *of course I did poorly on these tests.* They required me to keep a series of recently learned concepts in my head. A flag means "horse," a triangle means "boat," a square means "church." Wait — what does a flag mean again? Or the syllable three seconds ago was *mod,* the syllable two seconds ago was *er,* the syllable one second ago was *a,* and now the new syllable is *tion.* Hold on — what was that first syllable again? My success depended on my short-term memory, and (as is the case with many autistic people, I would later learn) my short-term memory is bad. So what else was new?

At the other extreme, I scored well at antonyms and synonyms because I could associate the test words with pictures in my mind. If the examining psychologist said "Stop" to me, I saw a stop sign. If he said "Go," I saw a green light. But not just any stop sign, and not just any green light. I saw a specific stop sign and a specific green light from my past. I saw a whole bunch of them. I even recalled a stop-and-go light from a Mexican customs station, a red light that turned green if the officers decided not to search your bags — and I'd seen that light more than ten years earlier.

Again: So what? As far as I could tell, everybody thought in pictures. I just happened to be better at it than most people, something I already knew. By this point in my life, I had been making architectural drawings for several years. I'd already had the experience of completing a drawing and looking at it and thinking, *I can't believe I did this!* What I hadn't thought was *I can do this kind of drawing because I have walked around the yard, committed every detail of it to memory, stored the images in my brain like a computer, then retrieved the appropriate images at will. I can do this kind of drawing because I'm a person with autism.* Just as I didn't think, *I scored in the sixth percentile in reasoning and in the ninety-fifth percentile in verbal ability because I'm a person with autism.* And the reason I didn't think these thoughts was that "person with autism" was a category that was only then beginning to come into existence.

Of course, the word *autism* had been part of the psychiatric lexicon since 1943, so the idea of people having autism had been around at least as long. But the definition was loose, to say the least. Unless someone pointed out an oddity in my behavior, I simply didn't go around thinking of what I was doing in terms of my being a person with autism. And I doubt that I was the exception in this regard.

The second edition of the *Diagnostic and Statistical Manual of Mental Disorders* was published in 1968, and, unlike its 1952 predecessor, it contained not one mention of autism. As best as I can tell, the word *autistic* did appear twice, but again, as in the *DSM-I,* it was there only to describe symptoms of schizophrenia and not in connection with a diagnosis of its own. "Autistic, atypical, and withdrawn behavior," read one reference; "autistic thinking," read another.

In the 1970s, however, the profession of psychiatry went through a complete reversal in its way of thinking. Instead of looking for causes in the old psychoanalytic way, psychiatrists began focusing on ef-

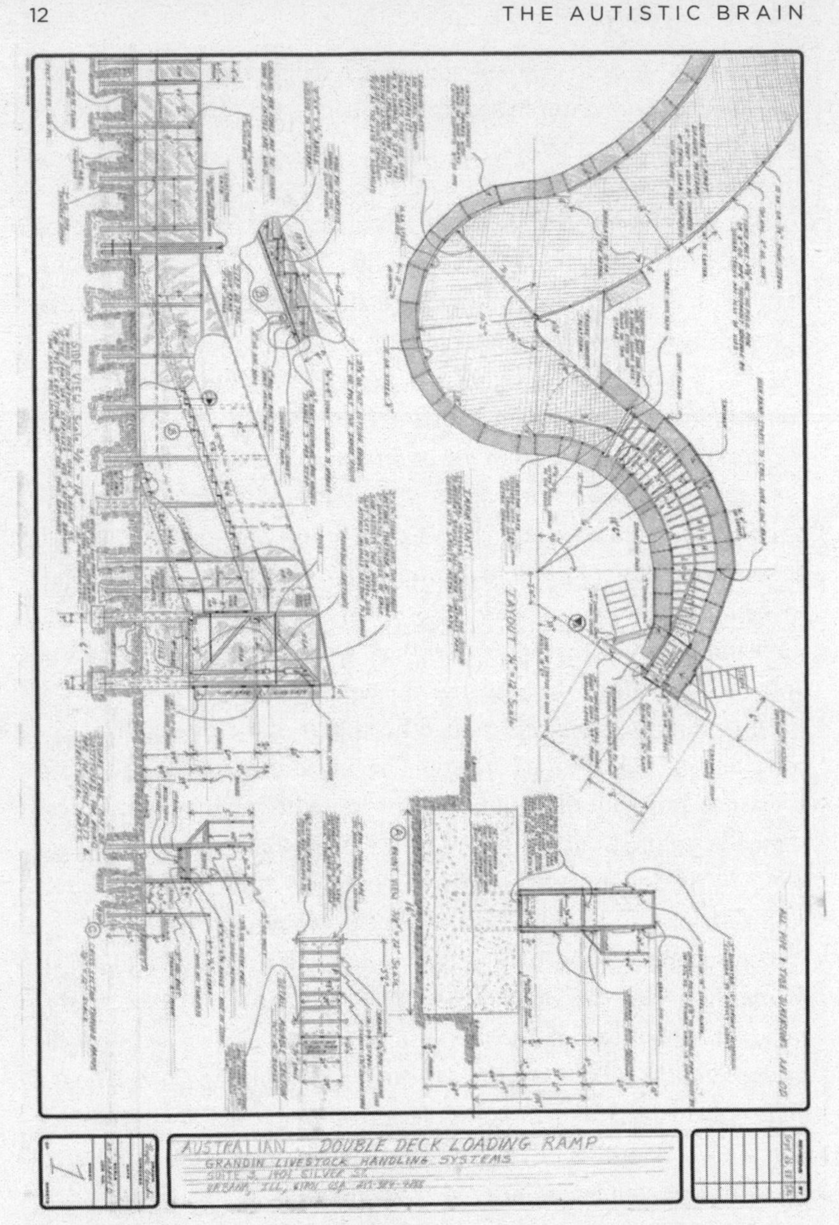

Being able to "download" images from my visits to cattle-handling facilities in order to create this blueprint for a double-deck loading ramp didn't seem unusual to me. © *Temple Grandin*

fects. Instead of regarding the precise diagnosis as a matter of secondary concern, the profession began trying to classify symptoms in a rigid and orderly and uniform fashion. The time had come, psychiatrists decided, for psychiatry to become a science.

This reversal happened for a few reasons. In 1973 David Rosenhan, a Stanford psychiatrist, published a paper recounting how he and several colleagues had posed as schizophrenics and fooled psychiatrists so thoroughly that the psychiatrists actually institutionalized them, keeping them in mental hospitals against their will. How scientifically credible can a medical specialization be if its practitioners can so easily make incorrect diagnoses — misdiagnoses, moreover, with potentially tragic consequences?

Another reason for the reversal was sociological. In 1972, the gay rights movement protested the *DSM*'s classification of homosexuality as a mental illness — as something that needed to be cured. They won that battle, raising the question of just how trustworthy *any* diagnosis in the *DSM* was.

But probably the greatest factor in changing the focus of psychiatry from causes to effects, from a search for a psychic injury to the cataloging of symptoms, was the rise of medication. Psychiatrists found that they didn't need to seek out causes for symptoms to treat patients. They could ease a patient's suffering just by treating the effects.

In order to treat the effects, however, they had to know what medications matched what ailments, which meant that they had to know what the ailments were, which meant that they were going to have to identify the ailments in a specific and consistent manner.

One result of this more rigorous approach was that the APA task force finally asked the obvious question: What is this autistic behavior that is a symptom of schizophrenia? In order to answer the question, the task force had to isolate autistic behavior from the other symptoms suggesting schizophrenia (delusions, hallucinations, and so on). But in order to describe autistic behavior, they had to describe autistic *behaviors*—in other words, have a checklist of symptoms. And a checklist of symptoms that didn't overlap with the other symptoms of schizophrenia suggested the possibility of a separate diagnosis: infantile autism, or Kanner's syndrome.

The *DSM-III*, published in 1980, listed infantile autism in a larger category called pervasive developmental disorders (PDD). To receive a diagnosis of infantile autism, a patient had to meet six criteria. One of the them was an absence of symptoms suggesting schizophrenia. The others were:

- Onset before 30 months
- Pervasive lack of responsiveness to other people
- Gross deficits in language development
- If speech is present, peculiar speech patterns such as immediate and delayed echolalia, metaphorical language, pronominal reversal
- Bizarre responses to various aspects of the environment, e.g., resistance to change, peculiar interests in or attachments to animate or inanimate objects

But that description was hardly precise. In fact, it became something of a moving target, changing with each new edition of the *DSM* as the APA attempted to nail down precisely what autism was—a common enough trajectory in psychiatric diagnoses that depend on observations of behavior. In 1987, the revision to the *DSM-III*, the *DSM-III-R*, not only changed the name of the diagnosis (from infantile autism to autistic disorder) but expanded the number of diagnostic criteria from six to sixteen, divided them into three categories, and specified that a subject needed to exhibit at least eight symptoms total, with at least two coming from category A, one from category B, and one from category C. This Chinese-menu sensibility led to higher rates of diagnosis. A 1996 study compared the *DSM-III* and *DSM-III-R* criteria as they applied to a sample of 194 preschoolers "with salient social impairment." According to the *DSM-III,* 51 percent of the children were autistic. According to the *DSM-III-R,* 91 percent of *the same children* were autistic.

The 1987 edition of the *DSM* also expanded an earlier diagnosis in the PDD category, atypical pervasive developmental disorder, into a catchall diagnosis that covered cases in which the symptoms of autism were milder or in which most but not all symptoms were present: pervasive developmental disorder not otherwise specified (PDD-NOS). The *DSM-IV,* which was published in 1994, further complicated the definition of autism by adding a new diagnosis altogether: Asperger syndrome.

In 1981, the British psychiatrist and physician Lorna Wing had introduced to English-language audiences the work Austrian pediatrician Hans Asperger had done in 1943 and 1944. Even as Kanner was trying to define autism, Asperger was identifying a class of children who shared several distinct behaviors: "a lack of empathy, little ability to form friendships, one-sided conversations, intense absorption in a special interest, and clumsy movements." He also noted that these children could talk endlessly about their favorite subjects; he dubbed them "little professors." Asperger called the syndrome "autistic psychopathy," but Wing felt that because of the unfortunate associations that had attached to the word *psychopathy* over the years, "the neutral term Asperger syndrome is to be preferred."

This addition to the *DSM* is important in two ways. The obvious one is that it gave Asperger's formal recognition by the psychiatric authorities. But when taken together with the PDD-NOS and its autistic-symptoms-but-not-quite-autism diagnostic criteria, Asperger's was also meaningful in how it changed the way we think about autism in general.

The inclusion of autism in the *DSM-III* in 1980 was significant for formalizing autism as a diagnosis, while the creation of PDD-NOS in the *DSM-III-R* in 1987 and the inclusion of Asperger's in the *DSM-IV* in 1994 were significant for reframing autism as a spectrum. Asperger syndrome wasn't technically a form of autism, according to the *DSM-IV*; it was one of five disorders listed as a PDD, alongside autism disorder, PDD-NOS, Rett syndrome, and childhood disintegrative disorder. But it quickly gained a reputation as "high-functioning autism," and by the time the revision of the *DSM-IV* appeared in 2000, diagnosticians were using *pervasive developmental disorder* and *autism spectrum disorder* (or ASD) interchangeably. At one end of the spectrum, you might find the severely disabled. At the other end, you might encounter an Einstein or a Steve Jobs.

That range, though, is part of the problem. It was almost certainly no coincidence that just as the idea of an autism spectrum was entering the mainstream of both popular and medical thinking, so was the concept of an autism "epidemic." If the medical community is given a new diagnosis to assign to a range of familiar behaviors, then of course the incidence of that diagnosis is going to go up.

Did it? If so, wouldn't we see a drop in some other diagnoses — the diagnoses that these new cases of autism or Asperger's would have previously received?

Yes — and in fact, we do see evidence of that drop. In the United Kingdom, some of the symptoms of autism would have previously been identified as symptoms of speech/language disorders, and those diagnoses in the 1990s did go down in roughly the same proportion that autism diagnoses went up. In the United States, those same symptoms would have received a diagnosis such as mental retardation, and, again, the number of those diagnoses went down as autism diagnoses went up. A Columbia University study of 7,003 children in California diagnosed with autism between 1992 and 2005 found that 631, or approximately one in eleven, had had their diagnoses changed from mental retardation to autism. When the researchers factored in those subjects who hadn't previously been diagnosed with anything, they found that the proportion of children who would have been diagnosed with mental retardation using older diagnostic criteria but who were now diagnosed with autism was *one in four.*

A later Columbia University analysis of the same sample population found that children living near autistic children had a greater chance of receiving the diagnosis themselves, possibly because their parents were more familiar with the symptoms. Is the kid talking on schedule? Does the child stiffen up and not want to be held? Can she play patty-cake right? Does he make eye contact? Not only were children who would once have been diagnosed with mental retardation now more likely to receive a diagnosis of autism, but more children were likely to receive a diagnosis of autism, period — enough to account for 16 percent of the increase in prevalence among that sample population.

I can see the effects of a heightened awareness of autism and Asperger's just by looking at the audiences who come to my talks. When I started giving lectures on autism in the 1980s, most of the audience members with autism were on the severe, nonverbal end of the spectrum. And those people do still show up. But far more common now are kids who are extremely shy and have sweaty hands, and I think, *Okay, they're sort of like me — on the spectrum but at the high-functioning end.* Would their parents have thought to have them tested for autism in the 1980s? Probably not. And then there are the geeky, nerdy kids I call Steve Jobs Juniors. I think back on kids I went to school with who were just like these kids but who didn't get a label. Now they would.

I recently spoke at a school for autistic students, to a hundred little kids sitting on the floor in a gymnasium. They weren't fidgeting much, so they were probably on the high-functioning end of the autism spectrum. But you never know. They looked to me just like the kids I had seen several months earlier at the Minnesota State Science Fair. Did the kids at the autism school get the diagnosis just so they could go to a school where they'd be left alone to do what they did best — science, history, whatever their fixations might be? Then again, did some of the kids at the science fair fit the diagnosis for autism or Asperger's?

The number of diagnoses of autism spectrum disorder almost certainly went up dramatically for another reason, one that hasn't gotten as much attention as it should: a typographical error. Shocking but true. In the *DSM-IV,* the description of pervasive developmental disorder not otherwise specified that was supposed to appear in print was "a severe and pervasive impairment in social interaction *and* in verbal or nonverbal communication skills" (emphasis added). What actually appeared, however, was "a severe and pervasive impairment of reciprocal social interaction *or* verbal and nonverbal communication skills" (emphasis added). Instead of needing to meet *both* criteria to merit the diagnosis of PDD-NOS, a patient needed to meet *either.*

We can't know how many doctors made an incorrect diagnosis of PDD-NOS based on this error. The language was corrected in 2000, in the *DSM-IV-TR.* Even so, we can't know how many doctors continued to make the incorrect diagnosis, if only because by then the incorrect diagnosis had become the standard diagnosis.

Put all these factors together — the loosened standards, the addition of Asperger's and PDD-NOS and ASD, the heightened awareness, the typographical error — and I would be surprised if there *hadn't* been an "epidemic."

I'm not saying that the incidence of autism hasn't actually increased over the years. Environmental factors seem to play a role in autism — *environmental* not only in the sense of toxins in the air or drugs in the mother's bloodstream, but other factors, like the father's age at the child's conception, which seems to affect the number of gene mutations in sperm, or the mother's weight during pregnancy. (See chapter 3.) If an environment changes for the worse — if a new drug comes on the market that we later discover causes autistic symptoms, or if a shift in the national work force leads more couples to wait to have children — the number of cases might rise. If an environment changes for the better — if services for children diagnosed with ASD become available in a community, prompting parents to doctor-shop until their kid gets the "right" diagnosis—well, the number of cases might rise then too.

For whatever combination of reasons, the reported incidence of autism diagnoses has only continued to increase. In 2000, the Centers for Disease Control and Prevention established the Autism and Developmental Disabilities Monitoring (ADDM) Network to collect data from eight-year-olds to provide estimates of autism and other developmental disabilities in the United States. The data from 2002 indicated that 1 in 150 children had an ASD. The data from 2006 raised the incidence to 1 in 110 children. The data from 2008—the most recent data available as I write this, and the basis for the most recent report, in March 2012—raised the incidence even further, to 1 in 88 children. That's a 70 percent increase in a six-year period.

The sample was 337,093 subjects in fourteen communities in as many states, or more than 8 percent of the nation's eight-year-olds that year. Given the size and breadth of that sample, the lack of geographical consistency was striking. The number of children identified with an ASD ranged erratically from one community to the next, from a low of 1 in 210 to a high of 1 in 47. In one community, 1 in 33 boys was identified as having an ASD. The rate of ASD incidence among black children was up by 91 percent from 2002. Among Hispanic children, the rise was even steeper—110 percent.

What's going on here? "At this point, it's not clear," Catherine Lord, the director of the Center for Autism and the Developing Brain in New York, wrote on CNN.com after the release of the 2012 report. And unfortunately, the *DSM-5*,[2] issued in 2013, doesn't clarify matters. (See chapter 5.)

You know how when you're cleaning out a closet, the mess reaches a point where it's even greater than when you started? We're at that point in the history of autism now. In some ways, our knowledge

2. The reason for the change from Roman numerals to Arabic is that Arabic numerals will allow easier updating: 5.1, 5.2, etc.

of autism has increased tremendously since the 1940s. But in other ways, we're just as confused as ever.

Fortunately, I think we're ready to pass that point of maximum confusion. As Jeffrey S. Anderson, the director of functional neuroimaging at the University of Utah School of Medicine, says, "There's a long tradition in medicine where the diseases start out in psychiatry and eventually they move into neurology"—epilepsy, for example. And now autism is joining that tradition. At long last, autism is yielding its secrets to the scrutiny of hard science, thanks to two new avenues of investigation that we'll explore in the next two chapters.

Over here, on the closet shelf corresponding to chapter 2, we'll put neuroimaging. Over there, on the shelf corresponding to chapter 3, we'll put genetics. We can begin to reorganize the closet with confidence, because now we have a new way of thinking about autism.

It's in your mind?

No.

It's in your brain.

2 Lighting Up the Autistic Brain

OVER THE YEARS, I've discovered I have a hidden talent. I'm very good at lying completely still for long periods of time.

The first time I realized I had this ability was in 1987, at the University of California, Santa Barbara, when I became one of the first autistic subjects to undergo magnetic resonance imaging, or MRI. The technicians warned me that the experience would be loud, which it was. They said the headrest would be uncomfortable, which it was. They said I had to lie very, very still, which, with some effort, I did.

None of these physical hardships, however, bothered me in the least. I was too excited. I was laying myself down on the altar of science! Slowly, my body slid into the big metal cylinder.

Not bad, I thought. *Sort of like the squeeze machine. Or something out of* Star Trek.

Over the following half an hour, everything I had been warned about happened: the sound of hammers on anvils; the crick in the neck; the self-conscious monotony of monitoring my every nonmovement. *Don't move, don't move, don't move*—thirty minutes' worth of telling myself to lie absolutely still.

And then it was over. I hopped off the gurney and headed straight for the technician's room, and there I received my reward: I got to see my brain.

"Journey to the center of my brain" is what I call this experience.

Seven or eight times now I have emerged from a brain-imaging device and looked at the inner workings that make me *me:* the folds and lobes and pathways that determine my thinking, my whole way of seeing the world. That first time I looked at an MRI of my brain, back in 1987, I immediately noticed that it wasn't symmetrical. A chamber on the left side of my brain — a ventricle — was obviously longer than the corresponding one on the right. The doctors told me this asymmetry wasn't significant and that, in fact, some asymmetry between the two halves of the brain is typical. But since then, scientists have learned how to measure this asymmetry with far greater precision than was possible in 1987, and we now know that a ventricle elongated to this extent seems to correlate with some of the symptoms that identify me as autistic. And scientists have been able to make that determination only because of extraordinary advances in neuroimaging technology and research.

Neuroimaging allows us to ask two fundamental questions about every part of the brain: What does it look like? What does it do?

Magnetic resonance imaging, or MRI, uses a powerful magnet and a short blast from a specific radio frequency to get the naturally spinning nuclei of hydrogen atoms in the body to behave in a way that the machine can detect. Structural MRI has been around since the 1970s, and as the word *structural* suggests, it provides views of the anatomical structures inside the brain. Structural MRI helps answer the What does it look like? question.

Functional MRI, which was introduced in 1991, shows the brain actually functioning in response to sensory stimuli (sight, sound, taste, touch, smell) or when a person is performing a task — problem-solving, listening to a story, pressing a button, and so on. By tracing the blood flow in the brain, fMRI presumably tracks neuron activity (because more activity requires more blood). The parts of the brain that light up while the brain responds to the stimuli or performs the assigned tasks, researchers assume, provide the answer to the What does it do? question. Over the past couple of decades, neurological research using fMRI studies has produced more than twenty thousand peer-reviewed articles. In recent years, that pace has accelerated to eight or more articles *per day.*

Even so, neuroimaging can't distinguish between cause and effect. Take one well-known example associated with autism: facial recognition. Neuroimaging studies over the decades have repeatedly indicated that the cortex of an autistic doesn't respond to faces as animatedly as it does to objects. Does cortical activation in response to faces atrophy in autistics because of the reduced social engagement with other individuals? Or do autistics have reduced social engagement with other individuals because the connections in the cortex don't register faces strongly? We don't know.

Neuroimaging can't tell us everything. (See sidebar at the end of this chapter.) But it can tell us a lot. A technology that can look at a part of a brain and address What does it look like? and What does it do? can also answer a couple of bonus questions: How does the autistic brain look different from the normal brain? and What does the autistic brain do differently than the normal brain? Already autism researchers have been able to provide many answers to those two questions — answers that have allowed us to take the behaviors that have always been the basis of an ASD diagnosis and begin to match them to the biology of the brain. And as this new understanding of autism is harnessed to more and more advanced neuroimaging technologies, many researchers think that a diagnosis based in biology is not just feasible but near at hand — maybe only five years away.

I always tell my students, "If you want to figure out animal behavior, start at the brain and work your way out." The parts of the brain we share with other mammals evolved first — the primal emotional areas that tell us when to fight and when to flee. They're at the base of the brain, where it connects with the spinal cord. The areas that perform the functions that make us human evolved most recently — language, long-range planning, awareness of self. They're at the front of

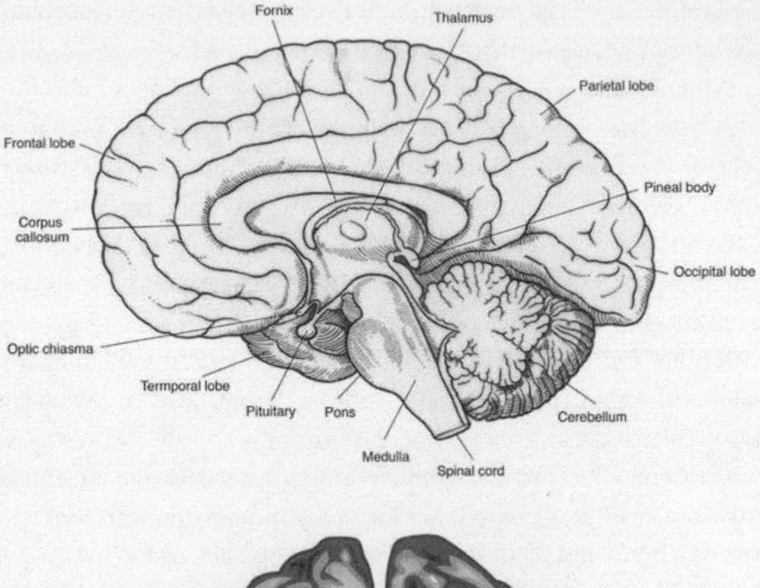

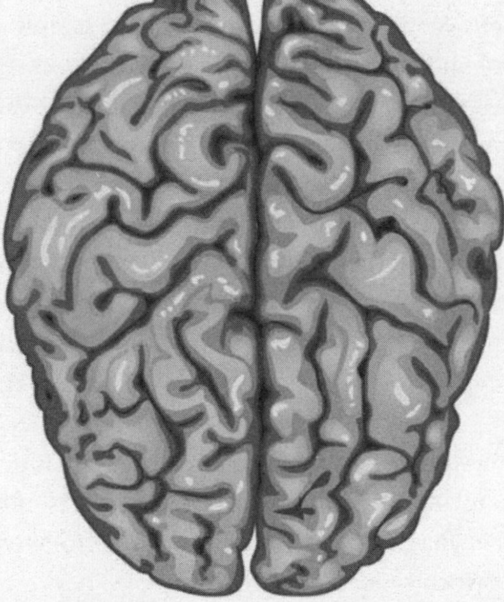

The human brain, side and overhead views.
© Science Source / Photo Researchers, Inc. (top); © 123rf.com (bottom)

the brain. But it's the overall complex relationship between the various parts of the brain that make us each who we are.

When I talk about the brain, I often use the analogy of an office building. The employees in different parts of the building have their own areas of specialization, but they work together. Some departments work closer together than others. Some departments are more active than others, depending on what the task at hand is. But at the end of the day, they come together to produce a single product: a thought, an action, a response.

At the top of the building sits the CEO, the prefrontal cortex—*prefrontal* because it resides in front of the frontal lobe, and *cortex* because it's part of the cerebral cortex, the several layers of gray matter that make up the outer surface of the brain. The prefrontal cortex coordinates the information from the other parts of the cortex so that they can work together and perform executive functions: multitasking, strategizing, inhibiting impulses, considering multiple sources of information, consolidating several options into one solution.

Occupying the floors just below the CEO are the other sections of the cerebral cortex. Each of these sections is responsible for the part of the brain it covers. You can think of the relationship between these discrete patches of gray matter and their corresponding parts as similar to the relationship between corporate vice presidents and their respective departments.

- The frontal cortex VP is responsible for the frontal lobe—the part of the brain that handles reasoning, goals, emotions, judgment, and voluntary muscle movements.
- The parietal cortex VP is responsible for the parietal lobe—the part of the brain that receives and processes sensory information and manipulates numbers.
- The occipital cortex VP is responsible for the occipital lobe—the part of the brain that processes visual information.
- The temporal cortex VP is responsible for the temporal lobe—

the auditory part of the brain that keeps track of time, rhythm, and language.

Below the VPs are the workers in these various divisions — the geeks, as I like to call them. They're the areas of the brain that contribute to specialized functions, like math, art, music, and language.

In the basement of the building are the manual laborers. They're the ones dealing with the life-support systems, like breathing and nervous system arousal.

Of course, all these departments and employees need to communicate with one another. So they have desktop computers, telephones, tablets, smartphones, and so on. When some folks want to talk to others face to face, they take the elevator or the stairs. All these means of access, connecting the workers in the various parts of the building in every way imaginable, are the white matter. Whereas the gray matter is the thin covering that controls discrete areas of the brain, the white matter — which makes up three-quarters of the brain — is a vast thicket of wiring that makes sure all the areas are communicating.

In the autistic brain, however, an elevator might not stop at the seventh floor. The phones in the accounting department might not work. The wireless signal in the lobby might be weak.

Before the invention of neuroimaging, researchers had to rely on postmortem examinations of the brain. Figuring out the anatomy of the brain — the answer to the What does it look like? question — was relatively straightforward: Cut it open, look at it, and label the parts. Figuring out the functions of those parts — the answer to the What does it do? question — was a lot trickier: Find someone who behaves oddly in life and then, when he or she dies, look for what's broken in the brain.

"Broken-brain" cases continue to be useful for neurology. Tumors. Head injuries. Strokes. If something's broken in the brain, you can really start to learn what the various parts do. The difference today, though, is that you don't have to wait for the brain's host to die. Neuroimaging allows us to look at the parts of the brain and see what's broken now, while the patient is still alive.

Once when I was visiting a college campus I met a student who told me that when he tried to read, the print jiggled. I asked him if he'd had any head injuries, and he said he'd been hit by a hockey puck. I asked where exactly he'd been hit. He pointed to the back of the head. (I don't think I was rude enough to actually feel the spot, but I can't say for sure.) The place where he was pointing was the primary visual cortex, which is precisely where I had expected him to point, because of what neuroimaging has taught us.

In broken-brain studies, we can take a symptom, an indication that something has gone haywire, and look for the wire or region that's damaged. Through this research, we have pinpointed the circuits in the back of the brain that regulate perception of shape, color, motion, and texture. We know which are which because when they're busted, weird stuff happens. Knock out your motion circuit, and you might see coffee pouring in a series of still images. Knock out your color circuit, and you might find yourself living in a black-and-white world.

Autistic brains aren't broken. My own brain isn't broken. My circuits aren't ripped apart. They just didn't grow properly. But because my brain has become fairly well known for its various peculiarities, autism researchers have contacted me over the years to ask permission to put me in this scanner or that. I'm usually happy to oblige. As a result of these studies, I've learned a lot about the inner workings of my own brain.

Thanks to a scan at the University of California, San Diego, School of Medicine's Autism Center of Excellence, I know that my cerebellum is 20 percent smaller than the norm. The cerebellum helps control motor coordination, so this abnormality probably explains why my sense of balance is lousy.

In 2006 I participated in a study at the Brain Imaging Research Center in Pittsburgh and underwent imaging with a functional MRI scanner and a version of MRI technology called diffusion tensor imaging, or DTI. While fMRI records regions in the brain that light up, DTI measures the movement of water molecules through the white-matter tracts — the interoffice communications among the regions.

- The fMRI portion of the study measured the activation in my ventral (or lower) visual cortex when I looked at drawings of faces and drawings of objects and buildings. A control subject and I responded similarly to the drawings of objects and buildings, but my brain showed a lot less activation in response to faces than hers did.
- The DTI scan examined the white-fiber tracts between various regions in my brain. The imaging indicated that I am overconnected, meaning that my inferior fronto-occipital fasciculus (IFOF) and inferior longitudinal fasciculus (ILF) — two white-fiber tracts that snake through the brain — have way more connections than usual. When I got the results of that study, I realized at once that they backed up something I'd been saying for a long time — that I must have an Internet trunk line, a direct line — into the visual cortex to explain my visual memory. I had thought I was being metaphorical, but I realized at that point that this description was a close approximation of what was actually going on inside my head. I went looking for broken-brain studies to see what else I could learn about this trunk line, and I found one that involved a forty-seven-year-old woman with visual memory disturbance. A DTI scan of her brain revealed that she had a partial disconnection in her ILF. The researchers concluded that the ILF must be "highly involved" in visual memory. *Boy,* I remember thinking, *break this circuit and I'm going to be completely messed up.*

In 2010 I underwent a series of MRI scans at the University of Utah. One finding was particularly gratifying. Remember that when I pointed out the size difference in my ventricles to the researchers after my first MRI, back in 1987, they told me that some asymmetry in the brain was to be expected? Well, the University of Utah study showed that my left ventricle is 57 percent longer than my right. That's huge. In the control subjects, the difference between left and right was only 15 percent.

My left ventricle is so long that it extends into my parietal cortex. And the parietal cortex is known to be associated with working memory. The disturbance to my parietal cortex could explain why I have trouble performing tasks that require me to follow several instructions in short order. The parietal cortex also seems to be associated with math skills — which might explain my problems with algebra.

Back in 1987, neuroimaging technology wasn't capable of measuring the anatomical structures within the brain with great precision. But if those researchers back then knew that one ventricle in my brain was 7,093 millimeters long while the other was 3,868 millimeters long, I guarantee it would have given them pause.

How did the two lateral ventricles become so different? One hypothesis is that when damage occurs early in the brain's development, other areas of the brain try to compensate. In my case, the damage would have occurred in the white matter in the left hemisphere, and the left ventricle would have enlarged to fill the damaged area. At the same time, the white matter in the right hemisphere would have tried to compensate for the lost brain function in the left hemisphere, and that expansion in the right hemisphere would have squeezed the right ventricle's growth.

The other significant findings from the Utah MRI study included:

- Both my intracranial volume — the amount of space inside the skull — and my brain size were 15 percent larger than the

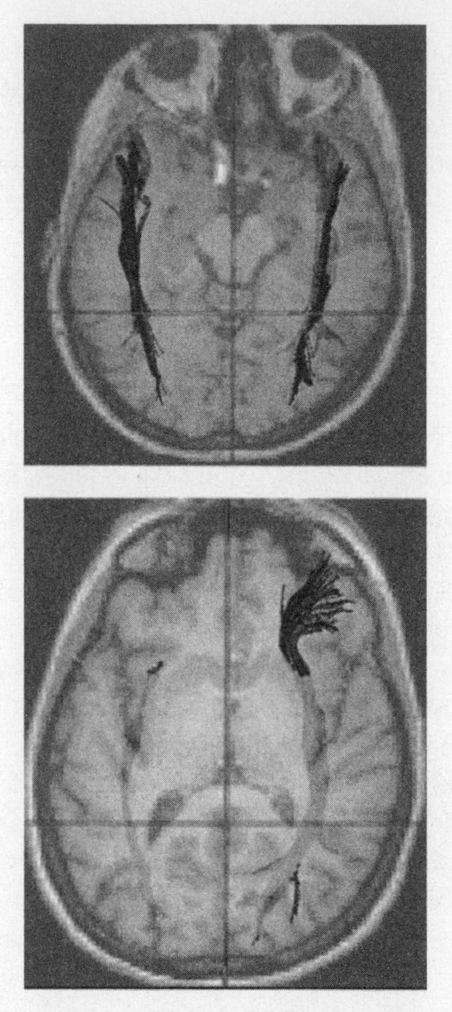

These scans from 2006 highlight (the areas in black from top to bottom) my inferior longitudinal fasiculus (ILF) and my inferior fronto-occipital fasciculus (IFOF). The ILF is much thicker than what a normal brain would show, and you can easily see how wildly my IFOF branches out. In both cases, these white-matter tracts stretch all the way back to the primary visual cortex, perhaps helping to explain my superb visual memory.

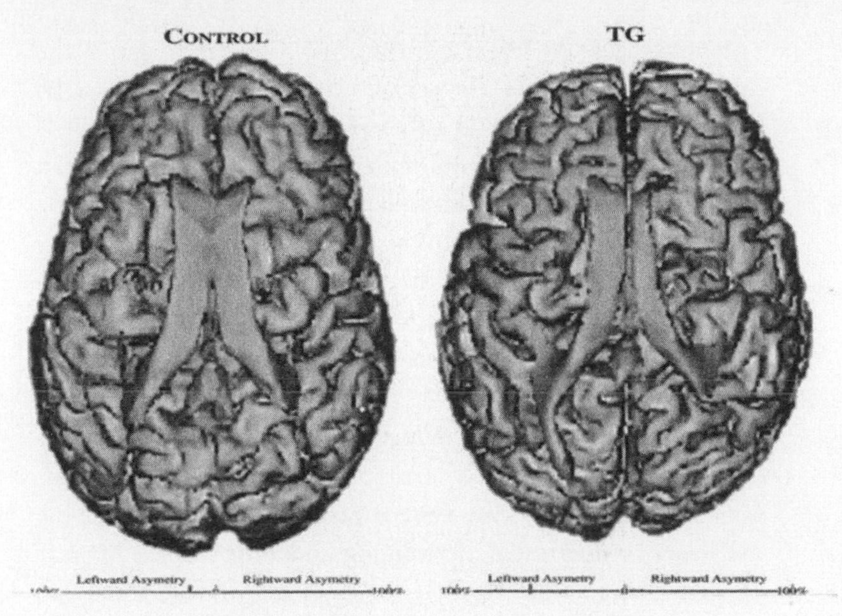

This scan from the University of Utah in 2010 dramatically shows that my left ventricle is much longer than my right — 57 percent longer. It's so long that it extends into the parietal cortex, an area associated with short-term memory, perhaps accounting for my poor ability at recalling several pieces of information in short order. *© Cooperrider, J.R. et al. presentation at the 2012 Society for Neuroscience meeting in New Orleans*

control subjects'. This too is likely the result of some sort of developmental abnormality. The neurons may have grown at an accelerated pace in order to compensate for the damaged area.

- The white matter in my left cerebral hemisphere was nearly 15 percent greater than the controls'. Again, this anomaly could be the result of an early developmental abnormality in my left hemisphere and my brain's attempt to compensate by generating new connections. This data reinforces for me the earlier University of Pittsburgh finding that my brain is overconnected.

- My amygdalae are larger than normal. The mean size of the three control subjects' amygdalae was 1,498 cubic millimeters. My left amygdala is 1,719 cubic millimeters, and my right is larger still — 1,829 cubic millimeters, or 22 percent greater than the norm. And since the amygdala is important for processing fear and other emotions, this large size might explain my lifelong anxiety. I think of all the panic attacks that plagued me through much of the 1970s, and they begin to make sense in a new way. My amygdalae are telling me I have everything to fear, including fear itself.

  Since I started taking antidepressants, in the early 1980s, the anxiety has been under control, probably because the pounding sympathetic nervous system reaction is blocked. But the vigilance is still present, percolating under the surface. My fear system is always on the alert for danger. If the students who live near me are talking in the parking lot under my window at night, I can't sleep. I actually turn on New Age music to block out the sound, even if the students are talking softly. (Though the music can't have vocals.) Volume has nothing to do with the fear factor; the association with a possible threat does. Human voices are associated with a possible threat. New Age music isn't associated with a possible threat. For that matter, neither is the sound of an airplane, so that sound doesn't bother me, even when I'm in a hotel by an airport. A plane could land on the hotel and I wouldn't wake up. But people talking in the next room? Forget it. I might as well turn on the light and read, because I know I'm not going to go to sleep until *they* go to sleep.
- The cortical thickness in both my left and right entorhinal cortices was significantly greater than the controls' — 12 percent in the left, and 23 percent in the right. "The entorhinal cortex is the golden gate to the brain's memory mainframe," says Itzhak Fried, a professor of neurosurgery at the David Geffen School of Medicine at UCLA. "Every visual and sensory experience that

we eventually commit to memory funnels through that doorway to the hippocampus. Our brain cells must send signals through this hub in order to form memories that we can later consciously recall." Maybe this peculiarity in my brain anatomy helps explain my exceptional memory abilities.

Naturally, I find these results fascinating because they highlight some of the odd things going on in my brain that help make me who I am. But what I find *really* fascinating is that they match the results of studies of some other people with autism.

- Preferring objects to faces? "These results are typical of individuals with autism," the researchers who conducted the MRI study at Pittsburgh in 2006 later wrote me in a summary of their findings. "One thing that seems to be coming up repeatedly in these scanning studies with individuals with autism is the marked reduction in the cortical activation to faces."
- Enlarged amygdalae are also often seen in people with autism. Because the amygdala houses so many emotional functions, an autistic can feel as if he or she is one big exposed nerve.
- And then there's this, in an e-mail from Jason Cooperrider, a graduate student who led the 2010 imaging study at Utah: "Dr. Grandin's head size is large by any standard, consistent with larger than average head/brain size/growth in autism." An enlarged brain can be caused by a number of genetic misfires, any one of which can result in an early spurt of neuronal development. The growth rate eventually normalizes, but the macrocephaly remains. The latest estimate is that about 20 percent of autistics have enlarged brains; the vast majority of those seem to be male, for reasons that aren't at all clear.

For the first time, thanks to hundreds if not thousands of neuroimaging studies of autistic subjects, we're seeing a solid match between autistic behaviors and brain functions. That's a huge deal. As one review article summarized the era, "This body of research clearly established autism and its signs and symptoms as being of neurologic origin." The long-held working hypothesis has now become the consensus of the evidence and the community: Autism really is in your brain.

The problem is, what's in *my* autistic brain is not necessarily what's in *someone else's* autistic brain. As the neuroanatomy pioneer Margaret Bauman once told me, "Just because your amygdala is larger than normal doesn't mean that every autistic person's amygdala is larger than normal." While some similarities among autistic brains have emerged, we have to be careful not to overgeneralize. In fact, neuroimaging researchers face three challenges to finding common ground among autistic brains.

Homogeneity of brain structures. While the 2010 Utah study revealed several striking anatomical anomalies in my brain, it also showed, as Cooperrider e-mailed me, that "for about 95% of the comparisons" with the control subjects, "the differences were negligible." This overwhelming normalcy in the autistic brain is the rule, not the exception.

"Anatomically, these kids are normal," Joy Hirsch, an autism researcher then at Columbia University Medical Center in New York, said regarding the subjects in a study of hers. "Structurally, the brain is normal on any scale that we can look at."

Which is not to say that the structures of the brains in her study, or autistic brains in general, don't vary from one brain to the next. They do. But that's true of normal brains too. It's just that the variations among the autistic brains predominantly fall within the range of what's normal. Thomas Insel, director of the National Institute of Mental Health, told *USA Today* in 2012, shortly after the Centers for Disease Control raised the estimated prevalence of autism from

1 in 110 to 1 in 88, "Even when you look at a child who has no language, who is self-injuring, who's had multiple seizures, you would be amazed at how normal their brains look. It's the most inconvenient truth about this condition."

Nonetheless, some patterns are emerging. In addition to the variations in my own brain that seem consistent with those of many other autistics — enlarged amygdalae, macrocephaly, lack of cortical engagement when looking at faces — these widespread patterns include:

- Avoiding eye contact. Different than a preference for objects over faces, this is the active avoidance of faces. A 2011 fMRI study in the *Journal of Autism and Developmental Disorders* found that the brains in a sample of high-functioning autistics and typically developing individuals seemed to respond to eye contact in opposite fashions. In the neurotypical brain, the right temporoparietal junction (TPJ) was active to direct gaze, while in the autistic subject, the TPJ was active to averted gaze. Researchers think that the TPJ is associated with social tasks that include judgments of others' mental states. The study found the opposite pattern in the left dorsolateral prefrontal cortex: in neurotypicals, activation to averted gaze; in autistics, activation to direct gaze. So it's not that autistics don't respond to eye contact, it's that their response is the opposite of neurotypicals'.

  "Sensitivity to gaze in dlPFC demonstrates that direct gaze does elicit a specific neural response in participants with autism," the study said. The problem, however, is "that this response may be similar to processing of averted gaze in typically developing participants." What a neurotypical person feels when someone won't make eye contact might be what a person with autism feels when someone *does* make eye contact.

And vice versa: What a neurotypical feels when someone does make eye contact might be what an autistic feels when someone *doesn't* make eye contact. For a person with autism who is trying to navigate a social situation, welcoming cues from a neurotypical might be interpreted as aversive cues. Up is down, and down is up.

- Overconnectivity and underconnectivity. A highly influential paper published in *Brain* in 2004 introduced an underconnectivity theory—the idea that underconnectivity between cortical regions might be a common finding in autism. On a global scale, the major sections of the brain can't coordinate their messages. Since then, numerous other studies have made the same argument, finding a relationship between underconnectivity between cortical areas and deficits in a variety of tasks related to social cognition, language, and executive function.

In contrast to this long-distance underconnectivity, other studies have found overconnectivity on a local scale. Presumably, this overgrowth occurs in ways I've already described, an attempt of one part of the brain to compensate for a deficit in another. The result can be positive. As I've mentioned, I exhibit overconnectivity in an area corresponding to visual memory. Fortunately I can manage the visuals. I can sit at a consulting session and run the movie in my mind of how a piece of equipment will work, and then I can turn it off when I'm done. Some people with autism, however, don't have an Off switch that works, and for them, overconnectivity leads to a barrage of information, much of it jumbled.

Which is not to say that the underconnectivity theory describes all autistic brains. Like many initial attempts to describe a solution to a problem, it probably oversimplifies the situation. As a 2012 study from the University of Amsterdam noted, "some patterns of abnormal functional connectivity in ASD are not captured by current theoretical models. Taken together, empirical findings measuring different forms of connectivity demonstrate complex patterns of abnormal connectivity in people with ASD." The theory, the paper concluded, "is in need of refinement."

Heterogeneity of causes. Even when researchers do think they've found a match between an autistic person's behavior and an anomaly in the brain, they can't be sure that someone else manifesting the same behavior would have the same anomaly. Part of the title of a 2009 autism study in the *Journal of Neurodevelopmental Disorders* captured the situation succinctly: "Same Behavior, Different Brains." In other words, just because you're prone to extreme anxiety doesn't mean your autistic brain has an enlarged amygdala.

Heterogeneity of behaviors. Conversely, when researchers find an anomaly in the brain, they can't be sure that that anomaly will have the same behavioral effect in a different brain. Or any effect, for that matter. Just because you have an enlarged amygdala doesn't mean that you're autistic.

But what if it did?

Not necessarily an enlarged amygdala. But what if some neuroanatomical finding or combination of them could serve as a reliable diagnostic tool? A diagnosis based not on behaviors alone but on biology as well would make a big difference in predicting deficits and targeting treatments. Doctors and researchers could:

- Apply early intervention, even in infancy, when the brain is still highly susceptible to being rewired.
- Target areas in the brain more locally, rehabilitating parts of the brain that they think they can help and not wasting time on parts that are unrecoverable.
- Test new therapies and monitor existing therapies more narrowly.

- Tailor a prognosis to an individual patient on a case-by-case basis.

For the patient, such a diagnosis would have a tremendous psychological benefit as well, by allowing him or her to know what's actually unusual. Personally, I *like* knowing that my high level of anxiety might be related to having an enlarged amygdala. That knowledge is important to me. It helps me keep the anxiety in perspective. I can remind myself that the problem isn't *out there* — the students in the parking lot under my bedroom window. The problem is *in here* — the way I'm wired. I can medicate for the anxiety somewhat, but I can't make it go away. So as long as I have to live with it, I can at least do so secure in the knowledge that the threat isn't real. The *feeling* of the threat is real — and that's a huge difference.

Given the obstacles to investigating autism from a neurological perspective — the homogeneity of brains, the heterogeneity of behaviors and causes — you might ask whether finding a biomarker is a realistic goal. Yet in recent years, researchers have made tremendous progress toward reaching that goal, and now many speak of *when*, not *if*.

"We still don't have a litmus test for autism," the neuroscientist Joy Hirsch said. "But we have a basis for it."

As the director of the Functional MRI Research Center at the Columbia University Medical Center in New York City, Hirsch has tried to build that foundation in the search for a litmus test. In a study her group conducted between 2008 and 2010, fifteen autistic subjects ranging in age from seven to twenty-two and twelve control children ranging from four to seventeen underwent fMRI scans of the superior temporal gyrus — the part of the auditory system that processes the sounds of speech into meaningful language. "The most obvious disability in autism is the disability of speech," she said, regarding the rationale behind the experiment. "Our hypothesis was that at the first stage we could begin to see differences." And they felt they did: Their measures of activity in that region could identify fourteen out of fifteen of the autistic subjects, a sensitivity rate of 92 percent. (Other researchers have questioned the reliability of comparing subjects who were awake and subjects who were sedated — factors that Hirsch's team felt they accounted for. As always in science, further tests will or will not reinforce the validity of the findings.)

Another way that research groups are searching for a biomarker is by taking a sample of autistic and control subjects, focusing on one aspect of the brain that the researchers have reason to believe is associated with autistic behavior, and seeing if they can create an algorithm that can tell one kind of brain from another. Jeffrey S. Anderson, from the University of Utah, offers this simplified description: "We use a whole bunch of normal brains and brains of individuals with autism, and we make a template of each one" — of autistic brain and neurotypical brain — "and we take a new subject in and just ask, 'Well, which one does it match more?'"

The point isn't to identify this brain or that brain as belonging to an autistic person or a neurotypical. It's to find an aggregate that could help identify areas of potential interest that might be biomarkers.

In a major study that Anderson's group published in 2011, the aspect of the brain under consideration was connectivity. The earlier studies indicating that autistic brains tend to have local overconnectivity and long-distance underconnectivity had focused on a small number of discrete brain regions. Anderson and his colleagues instead studied the connectivity of the entirety of the gray matter. Using a variation of fMRI called functional connectivity MRI, they obtained connectivity measurements among 7,266 "regions of interest." In a group of forty male adolescents and young adults with autism and a like sample of forty typically developing subjects, Anderson found that the connectivity test could identify whether a brain was autistic or typical with 79 percent accuracy overall and 89 percent accuracy for subjects who were under the age of twenty.

That level of accuracy is consistent with results from other research groups. A 2011 MRI study from the University of Louisville found that in a sample of seventeen autistic and seventeen neurotypical subjects, the length of the centerline of the corpus callosum could be used to distinguish between the two types of brains with a level of accuracy ranging from 82 percent to 94 percent, depending on statistical confidence levels.

In another MRI study from 2011, researchers at the Stanford University School of Medicine and Lucille Packard Children's Hospital looked not at the size of an individual part of the brain, as structural MRI studies usually do, but at the topology of the gray matter's folds — the brain's cliffs and valleys. In a sample of twenty-four autistic children and twenty-four typically developing children (all aged eight to eighteen), they identified differences between the two groups in the default mode network, a system associated with daydreaming and other brain-at-rest, nontask activities. The study subjects whose brains showed the greatest deviations from the norm also exhibited the most severe communication deficits. Volume measurements of the posterior cingulate cortex in particular achieved an accuracy rate of 92 percent in telling one kind of brain from the other.

Accuracy rates in the 80 to 90 percent range are not high enough for researchers to claim they've discovered a marker for autism, but it's progress of a sort that would have been difficult to imagine only a decade ago. And it's certainly high enough to inspire confidence in the algorithmic approach.

One of the goals for further research is to adapt these techniques to younger subjects. As Utah's Anderson says, "It's not really helpful to diagnose a teenager with autism, because we already know it." The younger the subject, the earlier the possibility of intervention. The earlier the intervention, the greater the potential effect on the trajectory of an autistic person's life.

Just how young a person in the scanner can be depends in part on the technology. Functional MRI, for instance, requires responses to stimuli that create brain activity, so children need to be old enough (and, of course, to possess the neurological capacity) to understand the stimuli. Structural MRI, including DTI, doesn't rely on brain activity, so it allows researchers to study subjects who are even younger — so young, in fact, that they might not exhibit behavioral signs of autism yet.

That was the case in a 2012 DTI study led by researchers from the University of North Carolina at Chapel Hill. The participants were ninety-two infants who all had older siblings diagnosed as autistic and therefore were thought to be at high risk themselves. Researchers scanned the subjects' brains at six months, then followed up with a behavioral assessment at twenty-four months (as well as further scanning in most cases). At that point, twenty-eight of the subjects in the study met behavioral criteria for ASD, and sixty-four did not. Did the white-matter fiber tracts of one group exhibit any differences from the tracts of the other group? Researchers concluded that in twelve of the fifteen tracts under investigation, they did. At the age of six months, the children who later developed autistic symptoms showed higher fractional anisotropy (or FA, the measure of the movement of water molecules through the white-matter tracts) than the rest of the children. Usually that would be a good sign; a higher FA indicates a stronger circuit. But by age twenty-four months, those same children were showing lower FA, a sign of a weaker circuit. Why were those same circuits stronger at six months than those of the children who were developing typically? Were they even stronger even earlier? The researchers don't have an answer, but they do have a new goal: three-month-olds.

Another goal for further research is to look at the brain in even finer detail. Fortunately, the future is already here. I know, because I've seen it.

Actually, I've been *inside* the future — a radically new version of DTI called high-definition fiber tracking. HDFT was developed at the Learning Research and Development Center at the University of

Pittsburgh. Walter Schneider, senior scientist at the center, explains that HDFT was underwritten by the Department of Defense to investigate traumatic brain injuries: "They came to me saying, we need something that can do for brain injury what X-rays do for orthopedic injury."

When the research team posted a paper on the *Journal of Neurosurgery*'s website in March of 2012, the technology got a fair amount of media attention. The paper reported on the case of a thirty-two-year-old male who had sustained a severe brain injury in an all-terrain-vehicle accident. (No, he wasn't wearing a helmet.) HDFT scans revealed the presence and location of fiber loss so precisely that the research team accurately predicted the nature of the lasting motor deficit — severe left-hand weakness — "when other standard clinical modalities did not."

"Just like there are 206 bones in your body, there are major cables in your brain," Schneider says. "You can ask most anybody on the street to create a drawing of what a broken bone looks like, and they would be able to draw something somewhat sensible. If you ask them, 'So what does a broken brain look like?' most people — including researchers in the field — can't give you the details."

Including researchers in the field? Really?

"A fuzzy image of bones doesn't give you a clean diagnosis," Schneider says. "We took diffusion tensor imaging, and made it so it can."

While the focus of HDFT research so far has been on traumatic brain injuries, Schneider's long-range plan is to map the information superhighways of the brain. For years I've compared the circuitry of the brain to highways, and I'm hardly alone. But the *high-definition* part of HDFT technology has revealed just how apt the superhighways reference is.

Regular DTI technology shows the highways and off-ramps and crossroads of your brain as if they were all on a two-dimensional map. That kind of map is useful if you want to know whether a fiber gets from here to there. It can show you where I-94 and County

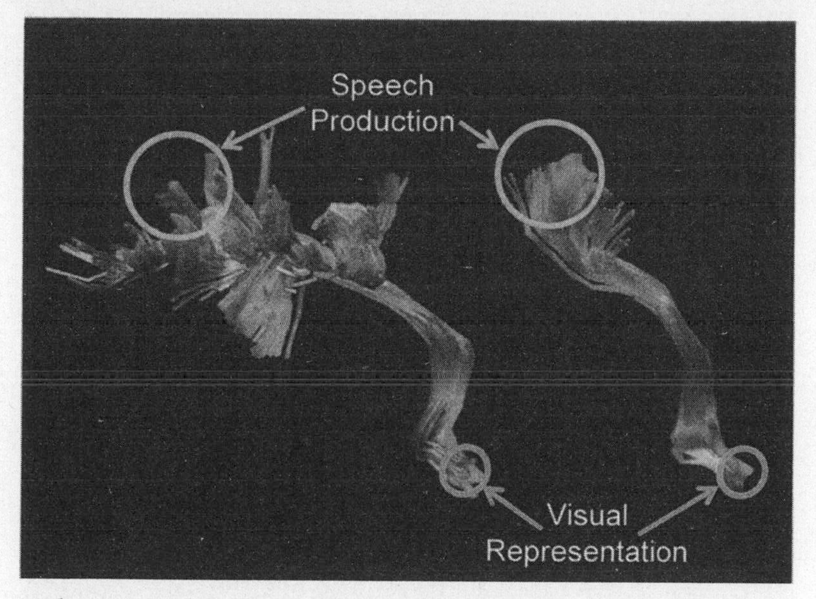

My brain on High Definition Fiber Tracking (left). Not only does HDFT reveal how disorganized my speech production and visual representation areas are compared to the control's, but it shows the fibers in unprecedented and glorious detail. © *Walter Schneider*

Road 45 are in close proximity to each other. It can show you that they crisscross. But it can't show you how they crisscross. Do they intersect, like a crossroads? Or does one road go over the other, like an overpass? The old technology can't answer that question. HDFT can.

And it *tracks* the fibers. It keeps them individualized over long stretches.

And it tracks the fibers *farther* than any previous technology — all the way to the end of the road.

It even shows if a damaged circuit still has continuity or if it's stopped transmitting. (As a biologist, I'm just freaking out, it's so cool.)

I don't want to overhype HDFT. It's incredibly important, but it's not going to solve all the mysteries of the brain. As Schneider says, "One of my favorite lines of neuroscience is if you can think of five ways for the brain to do something, it does it in all ten. The five you've thought of, and the five you haven't thought of yet." Still, HDFT is going to have a major impact on diagnoses involving brain trauma.

First, the diagnoses are going to be more precise. The existing state-of-the-art DTI scanner collects data from 51 directions. HDFT collects data from 257 directions. As a result, HDFT doesn't just tell you what section of the brain has been damaged. It tells you what specific fibers have been damaged, and how many.

Second, the diagnoses are going to be more persuasive. You know how athletes sometimes collapse and die? Everybody makes the connection between cause and effect—between overexertion and a strain on the heart—because the tragedy is visible and vivid and immediate. There's no mistaking it. And then the autopsy comes back, and it's unambiguous. The high-school football player died of a heart attack. The college basketball player died of a coronary aneurysm. But brain injuries have lacked a similar sort of clarity and immediacy, and therefore they've also lacked a similar sort of urgency. When a football player suffers a concussion or when a boxer takes multiple punches to the head, the effects of an injury might not be evident for years or decades. Not anymore. HDFT will show what the blows to the head have done to the brain, and I'm telling you, it's not going to be pretty. You won't need a medical degree to compare a concussed brain and a control brain and go, "Oh *no.*"

"In the case of brain trauma," Schneider says, "we're looking at a break in one of these cables." Not so in autism. There, he said, "we're looking at an anomalous growth pattern, be it genetic, be it developmental, et cetera, within that process."

I was invited to Schneider's lab to be scanned as part of a television program. Afterward, Schneider explained to me that he had been looking for areas in my brain that showed at least a 50 percent difference from the corresponding areas in a control subject. Two findings, he said, "really jumped out."

One, my visual tract is huge—400 percent of a control subject's.

Two, the "say what you see" connection in the auditory system is puny—1 percent of a control subject's. This finding made sense. In my book *Emergence,* I discussed my childhood speech problem: "It was similar to stuttering. The words just wouldn't come out."

I later asked Schneider to interpret these findings for me. Because we're still figuring out the brain, his interpretation would need to be in the nature of a hypothesis. But that's how science works. You gather information (my brain scans), use it to formulate a hypothesis, and make a prediction you can verify.

Between birth and the age of one, Schneider explained, infants engage in two activities that developmental researchers call verbal babbling and motor babbling. Verbal babbling refers to the familiar act of babies making noises to hear what they sound like. Similarly, motor babbling refers to actions such as waving a hand just to watch it move. During this period when babies are figuring out how to engage with the world, their brains are actually building connections to make that engagement possible. During verbal babbling, fibers are growing to make the connection between the "what you're hearing" and "what you're saying" parts of the brain. During motor babbling, fibers are growing to make the connection between the "what you're seeing" and "what you're doing" parts of the brain.

Then between the ages of one and two, children reach a stage where they can say single words. What's happening in the child's brain at this point is that fibers are forming an interlink between those two fiber systems that were constructed during the verbal and motor babbling period. The brain is connecting "what you're seeing" with "what you're saying" until out pops *Mama, Dada, ball,* and so on.

In my case, Schneider hypothesized, something happened developmentally during the single-word phase so that the fibers didn't form a connection between "what you're seeing" and "what you're saying." This would be the tract that was 1 percent of the size of the control subject's. To compensate, my brain sprouted new fibers, and they tried to go somewhere, anywhere. Where they wound up primarily was in the visual area rather than traditional language-production areas. That's the tract that was 400 percent of the size of the control subject's.

In such a scenario, Schneider continued, the babbling phase might be normal but language development would slow down dramatically between ages one and two.

Which would match a developmental pattern that the parents of children diagnosed with autism often report.

"Exactly," Schneider said.

But, he emphasized, the scenario he described was still only a hypothesis. He'll need more data, more scans that actually reflect how brains grow. "We've never had the technology to measure that," he said. "The project I'm working on is to map that developmental sequence."

He hadn't planned to adapt the HDFT technology to map the development of the autistic brain, but a question from *60 Minutes* correspondent Lesley Stahl changed his mind. Schneider asked me for permission to show my scans to her for a segment on autism her show was preparing. (The original television program that had commissioned the scan never aired.) In order not to raise unrealistic hopes for desperate parents, Schneider wanted to mention that HDFT scanning to diagnose the autistic brain wasn't going to be available at a local hospital in the near future — that it would be at least five to ten years before even leading hospitals had access to this technology. Stahl let him. But here's how Schneider remembers her phrasing the question:

"So a mother with a four-year-old child who will be age fourteen before she gets a biological diagnosis of her child's brain damage — that delay would mean a decade or more of failed treatment attempts, lost ability to communicate and educate her child, and the emotional strain that accompanies an uncertain diagnosis. What might be done to speed that process up and to make it available in five years?"

"This," Schneider said, "is why I'm doing a project on autism."

Science often advances because of new developments in technology. Think of Galileo and the telescope. He was one of the first people to point a "tube of long seeing" at the night sky, and what he found there forever changed how we conceive of the universe: mountains on the moon, moons around Jupiter, phases of Venus, and far, far more stars than met the naked eye. The same is true of neuroimaging. You can think of it as a "mindoscope" (to borrow a coinage from Hirsch), an instrument with which we have just begun to explore the universe within and to gather preliminary answers to our questions about the autistic brain: How does it look different than a normal brain? and What does it do differently than a normal brain?

We now understand the biological connections between parts of the brain and many of the behaviors that make up the current diagnosis of autism. But we don't yet know the cause behind the biology—the answer to a third question: How did it get that way?

For that answer, we have to turn to genetics.

Neuroimaging isn't perfect. In order to understand and appreciate what it can do best, let's look at what it can and cannot do.

- An fMRI can't capture the brain's activity during the full range of human experience. By necessity, it can observe only the brain responses that a person can have while lying still for long periods.
- Neuroimaging also requires subjects to keep their heads still. In recent years, several studies reported that short-range connections in the brain weaken as children grow older, while long-range connections strengthen. Neuroscientists considered this news to be quite a significant advance in the understanding of the brain's maturation process. Unfortunately, a follow-up study by the authors of the original studies showed that the supposed changes in the brain's development disappeared once they took head movement into account. "It really, really, really sucks," the

lead investigator said. "My favorite result of the last five years is an artifact."

This finding didn't cause scientists to rethink every brain scan out there. But it did serve as an unambiguous warning about the need to take head movement into account. This caution applies especially to studies of people with autism and other neurodevelopmental disorders. Why? Because those subjects are precisely the ones who will have the most difficulty holding still. Researchers are racing to figure out a way to factor out head motion in neuroimaging studies, but even if they're successful, they will have to ask themselves whether the removal of data from studies of one group of subjects (like autistics) will skew comparisons with studies of neurotypical subjects.

Even if you do manage to hold still, you can still screw up a neuroimaging result—as I know from personal experience. During one fMRI study, I was shown a flight simulation. First I was swooping over the Grand Canyon. Then I was skimming over wheat fields. Then I was skipping over mountaintops. Then I was feeling sick—which didn't seem like a good idea when you're inside a scanner. So I closed my eyes. Whatever else that scan was, it sure wasn't perfect.

- Even the best neuroimaging is only as good as current technology. Neurons fire hundreds of impulses per second, but the signal itself takes several seconds to blossom, and then it lingers for tens of seconds. Temporally precise, it's not. And the resolution doesn't really capture activity at the level of the neuron itself. As an article in *Science* magazine said, "Using fMRI to spy on neurons is something like using Cold War–era satellites to spy on people: Only large-scale activity is visible."
- And there are the researchers themselves. They have to be careful how they interpret the results. For instance, they shouldn't assume that if a portion of the brain lights up, it's essential for the mental process being tested. In one study, researchers found that the hippocampus was ac-

tivated when subjects were performing a particular exercise, but researchers conducting another study found that lesions to the hippocampus didn't affect the subjects' ability to perform that same exercise. The hippocampus was indeed *part* of the brain's response, but it wasn't a *necessary* part of the response.

- Researchers also can't assume that if a patient is exhibiting abnormal behavior and the scientists find a lesion, they've found the source of the behavior. I remember sitting in a neurology lecture in graduate school and suspecting that linking a specific behavior with a specific lesion in the brain was wrong. I imagined myself opening the back of an old-fashioned television and starting to cut wires. If the picture went out, could I safely say I had found the "picture center"? No, because there were a lot of wires back there that I could cut that would make the TV screen go blank. I could cut the connection to the antenna, and the picture would disappear. Or I could cut the power supply, and the picture would disappear. Or I could simply pull the plug out of the wall! But would any of those parts of the television actually be the picture center? No, because the picture depends not on one specific cause but on a collection of causes, all interdependent. And this is precisely the conclusion that researchers in recent years have begun to reach about the brain—that a lot of functions depend on not just one specific source but large-scale networks.

So, if you ever hear that fMRI can tell us people's political preferences, or how they respond to advertising, or whether they're lying, don't believe it. Science is nowhere near that level of sophistication yet—and may never be.

3 Sequencing the Autistic Brain

ON SEPTEMBER 6, 2012, I was doing what I usually do when I need to kill time in an airport—lingering at a newsstand, flipping through magazines, browsing the front pages of newspapers—when a page 1 headline in the *New York Times* caught my eye: "Study Discovers Road Map of DNA." I grabbed the paper and read on: "The human genome is packed with at least four million gene switches that reside in bits of DNA that once were dismissed as 'junk' but that turn out to play critical roles in controlling how cells, organs and other tissues behave."

Well, it's about time, I thought. The idea of junk DNA had never made sense to me. I remember in graduate school hearing about junk DNA. I heard references to it in the classroom. I saw peer-reviewed research articles about it in *Science* and *Nature. Junk DNA* is not a nickname, even though it may sound like one; it is an actual scientific term. It's called junk DNA because, unlike the sequences of DNA that code for proteins, these sequences didn't seem to have any purpose.

That idea was ridiculous to me. The double helix had always reminded me of a computer program, and you would never write code that had a lot of unnecessary stuff. The "junk" *had to* serve a purpose. It had to be something like the gene's operating system. If you go into your computer and find a lot of weird files, you might wonder what they're for, but you wouldn't conclude that they served no purpose. And you sure wouldn't want to reverse a couple of zeros and ones just to see what happened. Same thing with junk DNA. If you messed around with it, the gene's "computer program" would not work.

I was hardly alone in harboring this deep suspicion. For years, scientists had been taking the idea of junk DNA less and less seriously. In fact, geneticists had started preferring the terms *noncoding DNA* and *dark matter,* both of which suggested that this kind of DNA was simply a mystery, not garbage. As I stood reading the article in the airport, I felt vindicated after so many years, which is always nice, but that's not what jumped out at me.

The article — amid many others that day and in the weeks to come that emphasized the non-junk DNA angle — was based on the results of a massive federal research effort called the Encyclopedia of DNA Elements, or Encode. The project involved 440 scientists from thirty-two laboratories around the world, and the group's first thirty papers had appeared a day earlier in *Nature, Genome Research,* and *Genome Biology.* In one common analogy, the earlier sequencing of the human genome by the Human Genome Project and by Craig Venter's Celera Genomics in 2001 "was like getting a picture of Earth from space," as one scientist told the *Times,* while Encode was like Google Maps: It told us "where the roads are," "what traffic is like at what time of the day," "where the good restaurants are, or the hospitals or the cities or the rivers." The Human Genome Project told us what the genome was. Encode has begun to tell us what it does.

But what really interested me was the article's explanation of how the genome does what it does. In order to appreciate its significance, you first have to understand what DNA looks like. We've all seen the popular image of the double helix: that corkscrew of seemingly endless combinations of A (adenine), C (cytosine), G (guanine), and T (thymine) bases. But that Tinker Toy model represents a strand of DNA that's stretched out. A strand of DNA completely unfurled would be about ten feet long. But it's not unfurled. Instead, DNA is so tightly coiled that it fits inside the microscopic cell nucleus. By looking at DNA in its natural state, Encode researchers found, as the *Times* reported, "that small segments of dark-matter DNA are often quite close to genes they control."

Now that, I thought, *is a mindblower.*

Until then, scientists had been thinking about DNA in its stretched-out form. But if you envision DNA as a tightly wound coil—and while I was standing in the airport, holding the *Times* in my hands, that's exactly what my picture brain was doing—then a noncoding piece of DNA could be flipping switches on coding DNA that's hundreds of thousands of base pairs away. In the stretched-out helix, they're nowhere near each other; in the coiled-up helix, they're adjacent to each other.

I couldn't wait to get my hands on my issue of *Nature.* After I got off the flight home, I drove straight to the post office, but the magazine hadn't arrived. I can't say I waited by the mailbox for the next few days, but as soon as it did arrive, I tore into it. The article "The Long-Range Interaction Landscape of Gene Promoters" was of special interest, and I particularly enjoyed the concluding sentence of its abstract: "Our results start to place genes and regulatory elements in three-dimensional context, revealing their functional relationships."

But after I'd finished devouring that issue of the magazine, I realized that the most important lesson wasn't in any one of the six Encode articles. It was, instead, in the overall impression that the articles made on me. Taken together, they helped me realize how much we don't know about genetics.

Like neuroimaging, the science of genetics is still in its infancy. In a hundred years, the state of our knowledge today will look primitive. Ask yourself what would happen if we sent a laptop and a flash drive back in time one hundred years. Would scientists be able to figure out how pictures are stored on a flash drive? Let's be generous and give them one hundred laptops, so they can do some destructive testing. What these scientists would do is get inside the flash drive and take the chip out. They would map the anatomy of the chip. They would give all the parts highfalutin but silly Latin names. (*Amygdala,* the name of the brain's emotion center? It's from the Latin word for "almond," because that's what it looks like. *Hippocampus,* the name of the brain's file finder? It's from the Latin word for "seahorse," for the same reason.) And these scientists would assume that all the parts put together are the Intel, because each PC has "Intel Inside" written on it. But these scientists would have absolutely no idea how the flash drive works.

That's pretty much where we are today with the brain and the genome.

For a scientist, that lack of knowledge is thrilling. A new field to explore! A chance to do fundamental, big-picture research, before the field gets really narrow and specialized! Questions that lead to other questions! What could be more fun?

For a parent waiting for answers about an autistic child *today,* however, the lack of knowledge can be extraordinarily frustrating.

Fortunately, we do have the beginnings of a body of knowledge about the genetics of autism. Even knowing that genetics plays a role in autism is a vast improvement on where we were only a few decades ago. It might be difficult to believe now, but whether DNA had anything to do with autism was open to question as late as 1977, when the first study of autism in twins was published. The sample size was small, but the results were nonetheless striking. The concordance rate — meaning that both twins share the trait — for infantile autism among pairs of identical twins was 36 percent (four sets of twins out of eleven total). But among ten pairs of fraternal twins, the concordance rate was zero. Both those numbers might seem low, but remember, this was three years before the *DSM-III* provided the first formal diagnostic criteria for autism. By today's diagnostic standards — our current definition of autism — the concordance rates in that same sample would be 82 percent (nine sets of twins out of eleven) for identical twins and 10 percent (one set out of ten) for fraternal twins. A follow-up study in 1995, using double the sample size, found a comparable result: 92 percent concordance rate for identical twins, and 10 percent for fraternal twins.

Because identical twins share the same DNA, these results strongly support the idea that the source of autism is genetic. But the influence of DNA is not absolute. If one identical twin has autism, the chance that the other one will have it too is very high. But it's not 100 percent. Why not?

Well, we could ask the same question about other subtle differences in identical twins. Their parents can always tell them apart, and in some cases the differences are obvious enough that anyone can tell them apart. One reason is that even when the *genotype*—the DNA at conception—is identical in both twins, the genes might work differently inside the cell. The other reason is that the genotypes might not be identical at birth, due to spontaneous mutations in the DNA of one or both of the twins. Both sets of genetic differences contribute to an individual's *phenotype*—the person's physical appearance, intellect, and personality.

Knowing that genetics plays a role in autism, of course, is only a start. The next question is, Which gene or genes?

Even into the early years of the twenty-first century, some researchers held out hope that autism might be the result of one or just a handful of gene deviations in an individual's DNA. Maybe autism was like Down syndrome, which, as researchers discovered in 1959, is directly attributable to an extra copy of chromosome 21—the first time that a copy number variation was recognized as a cause of intellectual disability. In the case of Down syndrome, the relationship between cause and effect is clear: This particular chromosome causes that particular syndrome. Geneticists have had some success in locating specific cause-and-effect genes in autism-related disorders. In Rett syndrome—a disorder of the nervous system that leads to developmental reversals that are often diagnosed as symptoms of autism—the cause is a defect in the gene for a particular protein,

MeCP2, located on the X chromosome. In tuberous sclerosis — a genetic disorder that causes tumors to grow and is accompanied by ASD in nearly half of all cases — changes in one of two genes, TSC1 and TSC2, are responsible. Fragile X syndrome — the most common cause of mental retardation in boys, and one that can lead to autism — is due to a change in the FMR1 gene on the X chromosome.

By and large, though, the genetics of autism isn't that simple. Nowhere near.

After the Human Genome Project and Celera Genomics mapped the human genome in 2001, dozens of institutions in nineteen countries banded together to form the Autism Genome Project, or AGP. Using a database of 1,400 families, these scientists deployed the gene chip, a new technology that worked at a much higher level of resolution than previous methods and that allowed them to look at thousands of DNA variants on a single chip all at once, rather than on a one-by-one basis. The researchers used this technology to look at each subject's entire genome — all twenty-three pairs of chromosomes — as well as particular areas that earlier research had pinpointed as possibly being of interest.

When phase one of the Autism Genome Project came to an end, in 2007, the consortium published a paper in *Nature Genetics* that did identify several specific areas of the genome as likely contributors to autism. Among the more promising avenues for further research is a mutation in the gene that codes for a protein called neurexin, which links directly with a protein called neuroligin to control how two brain cells connect across the synapse between them. During development, these interactions are crucial for directing neurons to their proper targets and for forming signaling pathways in the brain. This finding by the AGP reinforced earlier research indicating that mutations in the SHANK3 protein, which interacts with neuroligin protein at the synapse, are associated with an increased risk of ASD and mental retardation.

But in addition to serving as a direction for further research, the paper demonstrated the effectiveness of the strategy that AGP scientists had used to detect these mutations. They searched for copy number variations, or CNVs — submicroscopic duplications, deletions, or rearrangements of sections of DNA. These variations, which can vary in length and position on the chromosome, have the potential to disrupt gene function.

Where do these copy number variations come from? Most are inherited. At some point, an irregularity entered the gene pool, and it was passed down through the generations. But some CNVs aren't hereditary. They arise spontaneously, either in the egg or sperm before fertilization or in the fertilized egg shortly afterward. These are called de novo mutations, from the Latin words for "from the beginning."

Many CNVs are benign. And geneticists estimate that each genome — each person's unique DNA — might contain as many as several dozen de novo mutations. They're part of what makes each person unique. But might de novo CNVs be associated with autism?

This is the question that a 2007 study of 264 families, published in *Science,* set out to answer. The authors concluded that such mutations do pose "a more significant risk factor for ASD than previously recognized." The study found that 10 percent of autistic children with nonautistic siblings (12 out of 118) had de novo copy number variations, but only 1 percent of controls who had no history of autism (2 out of 196) showed CNVs. In the following five years, this paper, "Strong Association of De Novo Copy Number Mutations with Autism," would be cited more than 1,200 times.

The hope that autism could be traced to one or even a few gene variations became less and less realistic. By the time phase two of the Autism Genome Project — drawing on the DNA of 996 elementary-school-age children in the United States and Canada diagnosed with ASD, their parents, and 1,287 controls — came to an end, in 2010, the collaborators had identified dozens of copy number variants potentially associated with ASD. By 2012, geneticists had associated ASD with hundreds of copy number variations.

Further complicating the research was that many of the CNVs seemed to be, if not unique, at least extremely rare. The authors of the 2007 *Science* paper seeking to link de novo mutations with autism had noted: "None of the genomic variants we detected were observed more than twice in our sample, and most were seen but once." In 2010, upon the publication of the Autism Genome Project's phase-two research, UCLA professor of human genetics and psychiatry Stanley Nelson said, "We found many more disrupted genes in the autistic children than in the control group. But here's where it gets tricky — every child showed a different disturbance in a different gene." In September 2012, an article in *Science,* "The Emerging Biology of Autism Spectrum Disorders," recounted the stunning progress in the discovery of possible autism-related CNVs — but "with no single locus accounting for more than 1 percent of cases."

Geneticists sometimes speak of a many-to-one relationship: many candidate mutations, one outcome. But what outcome, specifically? A diagnosis of autism? A symptom of autism? As is the case in neuroimaging, trying to understand autism through genetics is complicated by its heterogeneity. Autism manifests itself in numerous traits, and those traits are not identical from individual to individual. Why should we expect that the genetics of autism would provide a one-to-one correspondence between mutation and diagnosis?

In fact, researchers are finding that some mutations can contribute to a range of diagnoses, including intellectual disability, epilepsy, ADHD, schizophrenia — a one-to-many relationship. Again, heterogeneity is the problem, because the diagnosis of autism is based on behaviors, and autism shares those behaviors with other diagnoses. If researchers knew which traits — if any — were specific to autism, the search for a genetic cause might be a lot easier. As G. Bradley Schaefer, a neurogeneticist at the Arkansas Children's Hospital Research

Institute, says, "The key is trying to figure out which differences are secondary versus which differences are salient to the condition."

Until they figure that out, researchers have to adopt other methodologies to pinpoint autism-related genes. The Autism Genome Project, for instance, looked for a pattern among the mutations, or at least the beginning of a pattern. And the researchers found it: Many of the genes belonged to categories known to affect cell proliferation and cell signaling in the brain — a pattern that further reinforced the previous findings about the significance of the neurexin-neuroligin linkage and SHANK3.

In 2012, three groups of researchers that had independently devised an identical new approach to discovering de novo mutations published their complementary findings in an issue of *Nature.* Their strategy was to include only autistic subjects whose parents and siblings exhibited no autistic behaviors. They then used letter-by-letter sequencing of the exome — the protein-coding parts of the genome — to identify de novo single-letter mutations. If they found a de novo CNV in at least two of their autistic subjects, and if that CNV did not appear in any of the nonautistic subjects, then they considered that mutation a contributing agent to autism.

One of those studies, led by Matthew W. State, a neurogeneticist at the Yale University School of Medicine's Child Study Center, sampled two hundred autistic children and their nonautistic parents and siblings and found two children with the same de novo mutation, one that none of the nonautistic participants showed. At the same time, another study, led by Evan E. Eichler at the University of Washington in Seattle, independently sampled 209 families and found a subject with the same de novo mutation as a subject in the Yale study. Again, it was one that neither study had found in their nonautistic subjects. The University of Washington study also identified another de novo CNV in two autistic participants in its own study. Then a third study, led by Mark J. Daly at Harvard, looked for those three de novo variations — the one from State's study, the one from Eichler's study, and the one the two studies shared — in a separate sample of subjects and identified children with autism who had the same CNVs, indicating a possible correlation between that CNV and autism.

Another finding from that same trio of studies is worth noting — CNVs were four times more likely to originate on the father's side than on the mother's. This finding received reinforcement a few months later with the publication of a paper in *Nature* that reported a correlation between a father's age and the rate of de novo mutations. For me, that paper was one of those "Of course!" slap-yourself-on-the-forehead moments. Sperm cells divide every fifteen days, more or less, so the older a father is, the greater the number of mutations in his sperm. It's like making a copy of a copy of a copy on a photocopier. And the greater the number of mutations, the higher the risk of a mutation that might contribute to autism.[1]

But even if geneticists do manage to correlate a mutation with autism (regardless of whether the mutation is related to other conditions), they still don't know if one mutation alone is sufficient to create an autistic-like trait, or whether the emergence of a single trait depends on a combination of mutations. In recent years, opinion has shifted toward this multiple-hit hypothesis, thanks in large part to findings coming out of Eichler's lab. "The development of the brain is probably very sensitive to dosage imbalances," he said, describing his findings. One insult — as geneticists call a mutation with the potential to damage health — may be enough to cause havoc. And two? Good luck.

That conclusion has been reinforced by other labs. For instance, a 2012 analysis of mutations in the SHANK2 gene — which codes for a synaptic protein, like SHANK3, neurexins, and neuroligins — would

1. On an individual basis, the increase in risk is extremely low. Only on a population-wide basis would a change in the incidence rate become statistically significant.

have been significant if it had found only further support for a link between autism and mutations in genes related to neural circuitry. But the study, based on 851 subjects diagnosed with ASD and 1,090 controls, also found that all three subjects with the de novo SHANK2 mutation also carried hereditary mutations in a section of chromosome 15 long associated with autism.

"For these patients, it's like the genome cannot cope with that extra de novo event," said lead investigator Thomas Bourgeron, professor of genetics at the University of Paris, Diderot. "It may be like nitro and glycerin. Alone they're okay. But if you mix the two, you have to be very careful."

For me, the multiple-hit hypothesis is supported by observations that I've made again and again when I've met with families over the past twenty years. I've noticed that in a lot of cases, a kid with autism has at least one parent who exhibits a mild form of autistic behavior. A kid with severe autism often has two parents who exhibit this behavior. If both parents are contributing copy number variations of a kind that pose a higher risk for autism, then the incidence of autism in the children in those families is naturally going to go up. The more you load the dice on both sides of the family, the likelier you are to have a kid with a problem.

So far I've been addressing only hereditary and de novo mutations — those that are present at or near conception. But geneticists also study what happens to genes throughout pregnancy and over the course of a lifetime — a period when environmental factors enter into consideration. Can automobile exhaust contribute to autism? The mother's diet during pregnancy? Vaccines?

If your genes carry a higher risk for an environmental factor triggering a disease or condition, then we would say you have a genetic *susceptibility* or *predisposition.* If environmental factors interact with your genes in such a way as to cause a genetic change, then we would say you have an *acquired* or *somatic* mutation. Research into environmental influences on autism, however, is much less conclusive, and often much more controversial, than research into genetic factors alone.

"It is widely accepted that autism spectrum disorders are the result of multiple factors, that it would be extremely rare to find someone who had a single cause for this behavioral syndrome," the environmental epidemiologist Irva Hertz-Picciotto said in 2011. "Nevertheless, previous work on genes has generally ignored the possibility that genes may act in concert with environmental exposures."

Hertz-Picciotto has served as the principal investigator of Childhood Autism Risks from Genetics and Environment (CHARGE), a research program at the Medical Investigation of Neurodevelopmental Disorders (MIND) Institute at the University of California, Davis. "We expect to find many, perhaps dozens, of environmental factors over the next few years," Hertz-Picciotto said, "with each of them probably contributing to a fraction of autism cases. It is highly likely that most of them operate in conjunction with other exposures and/or with genes."

What was the organizing principle behind such a massive project? Hertz-Picciotto says that from the start, the members of the collaboration had decided to divide their investigations into three areas: nutrition, air pollution, and pesticides.

The first CHARGE study to attract national attention, in the journal *Epidemiology* in 2011, found that the combination of certain unfavorable genes and a mother's lack of vitamin supplementation in the three months prior to conception and during the first month of pregnancy significantly increased the risk for autism. Another CHARGE study, published in 2011 in *Environmental Health Perspectives*, found that children born to mothers living less than two blocks from a freeway were more likely to have autism, presumably due to exposure to automotive exhaust. A third CHARGE study, published in 2012, found that among the mothers of children with ASD or developmental delays, over 20 percent were obese, while among the mothers of typically developing children, 14 percent were obese.

Some CHARGE studies have been much less conclusive — for instance, this finding from another 2012 paper: "Certain pesticides may be capable of inducing core features of autism, but little is known about the timing or dose, or which of various mechanisms is sufficient to induce this condition." In fact, the conclusion of that paper was essentially a plea for further research: "In animal studies, we encourage more research on gene × environment interactions, as well as experimental exposure to mixtures of compounds. Similarly, epidemiologic studies in humans with exceptionally high exposures can identify which pesticide classes are of greatest concern, and studies focused on gene × environment are needed to determine if there are susceptible subpopulations at greater risk from pesticide exposures." Direction for further research isn't unusual in scientific papers, but the breadth of the request in this case was notable. In fact, an editorial in the July 2012 issue of *Environmental Health Perspectives* made a similar plea — and not just regarding pesticides. Instead, it called for the investigation of anything out there that might be hazardous — the "formulation of a systematic strategy for discovery of potentially preventable environmental causes of autism and other NDDs," or neurodevelopmental disorders.

"I think people had unrealistic expectations," Hertz-Picciotto says. "People in the genetics field really thought that was going to be *the* story." Instead of "looking for the rarer and rarer and even rarer mutations," she says, they might have better luck trying to link environmental factors with *common* genetic variants.

I myself have often wondered if the increase in prescription-drug use over the past few decades has contributed to an increase in the incidence of autism. In June of 2011, the Food and Drug Administration issued a safety alert cautioning pregnant women about a possible connection between cognitive development and the use of valproate, a mood stabilizer as well as a seizure medication. Later that same year, two studies showed that children whose mothers had taken valproate during pregnancy had a higher risk for low IQ and other cognitive deficiencies as well as autism and other disorders along the ASD spectrum. "An estimated six to nine percent of babies exposed to valproate *in utero* develop autism," reported the website for the Simons Foundation Autism Research Initiative, "a risk several-fold higher than in the general population."

The first study to investigate a link between antidepressant use and autism specifically, conducted by the Kaiser Permanente Medical Care Program in northern California, didn't appear until 2011. The study compared 298 children with ASD, along with their mothers, to more than 1,500 control children, along with their mothers, and it did find evidence for a slightly higher risk among those mothers who used antidepressants during or immediately prior to pregnancy. Okay, I thought, but maybe a mother who needs antidepressants already has more at-risk CNVs, meaning that the trigger for autism might be something related to the depression, not to the antidepressants. But the study took that possibility into account and found that mothers who were depressed but did not take antidepressants showed no increased risk level.

Risk levels, though, are relative. The study concluded that antidepressants are "unlikely to be a major risk factor." But what about a minor risk factor? The research indicated that mothers who took antidepressants during the year before delivery had a 2.1 percent greater risk of having children with ASDs, and the greatest increase in risk, 2.3 percent, came when the drug was taken during the first trimester.

But here's the thing. I think Prozac is a fabulous drug. I have friends who would be in really bad shape if they weren't on Prozac, Lexapro, or some other selective serotonin reuptake inhibitor. I know people who have been *saved* by these drugs. I myself wouldn't be functional without them. They can transform a life merely being lived into a life worth living. So women who are pregnant or are thinking about becoming pregnant and who take antidepressants should consult a doctor and weigh the risks and benefits.

In any case, we have to be very careful about looking for cause-and-effect relationships between environmental factors and genetics. As every scientist knows, correlation does not imply causation. An observed correlation — two events happening around the same time — might just be coincidence. Let's use the now infamous vaccination controversy as a way to look at the logical complexity of a causation-versus-coincidence argument. The story goes like this.

Parents routinely have their children vaccinated around age eighteen months. Some parents note that their children begin exhibiting signs of autism around age eighteen months — withdrawing into themselves, reversing the gains they'd made in learning language, engaging in repetitive behaviors. Is the correlation between certain vaccines and the onset of autism an example of coincidence or causation? Along comes a study in the British journal *The Lancet* in 1998 that offers the answer: causation. Parental outrage ensues,[2] as does a widespread grassroots movement to persuade parents not to have their children vaccinated. Yet numerous follow-up investigations can't replicate the results of the 1998 study, and in 2010, following an investigation by the UK General Medical Council that determines the research was misleading and incorrect, *The Lancet* retracts the study.

End of story? Not quite.

In fact, some children have been known to get incredibly sick and manifest severe symptoms consistent with autism very shortly after receiving the eighteen-month vaccinations. In those rare cases, the correct diagnosis has turned out to be a mitochondrial disease. The nucleus of a cell holds the chromosomes; that's where our genes

---

2. Personally, I don't think people will consider the issue settled until someone runs a study that separates regressive subjects (those children who start out developing normally and then regress at around eighteen months) from nonregressive subjects.

are encoded. But outside of the nucleus, in the cytoplasm of the cell, are organelles (the word comes from the idea that organelles are to the cell what organs are to the organism), and some of these organelles are mitochondria. Every cell has hundreds to thousands of mitochondria. Their purpose is to take chemicals in the body and convert them into usable energy. Mitochondria have their own DNA, separate from the DNA in the chromosomes. And just like the DNA in the chromosomes, mitochondrial DNA can suffer mutations. In some cases, the vaccination and the onset of symptoms might indeed be related. Some of the symptoms might be relatively mild, some might be life-threatening, and some might include loss of muscle coordination, visual and hearing problems, learning disabilities, gastrointestinal disorders, neurological problems. All of these symptoms would be part of the mitochondrial disease, and all of them would be consistent with autism.

"There's intense research going on in this area," says G. Bradley Schaefer, a neurogeneticist at the Arkansas Children's Hospital Research Institute as well as the lead author of the guidelines for genetic testing in children for the American College of Medical Genetics in 2008. "But not enough is known to make conclusions." The 2013 update of the guidelines wasn't publicly available at the time of this writing, but Schaefer did summarize the recommendations in an interview for this book: "There's been this question about mitochondrial influence in autism, there's research going on, there's clearly anecdotal cases—but right now we don't recommend routine testing due to the lack of sufficient objective evidence to support it." (Also, such testing is expensive and difficult, and it usually requires a muscle biopsy.)

A perhaps more compelling example of a genetic predisposition is in the DRD4 gene, which codes for a receptor that regulates the level of dopamine in the brain. Some people possess a variant of the DRD4 gene called DRD4-7R, the *7R* for "7 repeat allele," meaning that its nucleotide sequence repeats seven times. The brains of people who possess the 7R version of the DRD4 gene are less sensitive to dopamine — a neurotransmitter that affects brain processes involving movement, emotional response, and the ability to experience pleasure and pain — putting them at risk for attention and conduct disorders. For this reason, the 7R version of DRD4 has been called the brat gene or the drinking gene.

On a more clinical (and linguistically charitable) level, numerous studies have linked this allele with anxiety, depression, epilepsy, dyslexia, ADHD, migraines, obsessive-compulsive behavior, and autism. For example, a study published in 2010 reported several associations between autistic children with the 7R variant and their parents.

- Children with the 7R variant who had at least one parent with the 7R variant were significantly more likely to exhibit tic-like behaviors than those whose parents did not have the 7R variant.
- If the father had the 7R variant, a child with the 7R variant was more likely to exhibit behavior consistent with obsessive-compulsive disorder and tic severity.
- If the mother had the 7R variant, a child with the 7R variant was more likely to exhibit behavior consistent with oppositional defiant disorder and social anxiety disorder.

Scientists have known for a while that children with the 7R version of DRD4 (as well as other "risk" genes, like MAOA and SERT) are vulnerable to negative influences from their environment — an abusive or unsupportive parent, for example. Those negative influences can produce more severe versions of whatever behavior the child is already manifesting. For this reason, scientists long considered the 7R version to be the "poster gene" for genes that interact with a negative environment to create negative behavior. Hence its nickname: vulnerability or risk gene.

But what if children with risk genes experienced parental affirmation or otherwise healthy home lives instead of bad environments?

While research was persuasive that negative environments tended to lead to negative behavior in people who had this variation of the DRD4 gene, what if that same research also contained data indicating that *positive* environments tended to lead to *positive* behavior — but because the researchers were trying to measure negative effects, they didn't ask the right questions?

Fortunately, other researchers did eventually think to ask. Once they began conducting studies specifically looking for positive effects — and reanalyzing older studies of negative effects — investigators realized that they needed to rethink how science saw these gene variations. People with these gene variations are simply more sensitive to their environments — "for better or worse," as one researcher said. You could think of them as "orchid children," because they easily flourish or wilt depending on whether the hothouse environment they inhabit is conducive to growth or not. By contrast, "dandelion children," who carry the regular version of the gene, fare just about the same no matter where they grow.

Under this new understanding of how the 7R version of DRD4 works, geneticists have begun referring to it not so much as a risk gene but as a *responsiveness* gene. Nature made it neutral. Nurture makes it positive or negative.

You might wonder if this interpretation means that Leo Kanner was right about the negative influence of negative parenting. Not quite. Kanner was drawing a one-to-one correspondence between a refrigerator parent and autism in the child. Bruno Bettelheim's version of Kanner's model at least considered the possibility of a genetic component — a genetic predisposition toward autism that needed an abusive parent in order to become manifest. But neither Kanner nor Bettelheim seems to have considered autism to be the result of genetic predetermination, rather than predisposition.

But you know who did? Despite all the discredited psychoanalytic associations embedded in Kanner's and Bettelheim's assumptions and hypotheses, the answer is Sigmund Freud — sort of.

Freud's medical background was in neurobiology and neuroanatomy. He always argued that his psychoanalytic concepts were placeholders until science could do better. "We must recollect that all of our provisional ideas in psychology will presumably one day be based on an organic substructure," he wrote in 1914. Six years later he continued that thought. "The deficiencies in our description would probably vanish if we were already in a position to replace the psychological terms by physiological or chemical ones," he wrote. "We may expect [physiology and chemistry] to give the most surprising information and we cannot guess what answers it will return in a few dozen years of questions we have put to it. They may be of a kind that will blow away the whole of our artificial structure of hypothesis."

The same is true today. Neuroimaging has allowed us to probe neuroanatomical features and ask the questions What does it look like? and What does it do? Genetics has allowed us to begin to answer the question How does the brain do what it does? While we have decades of progress ahead of us, we have at least begun to find a few of the answers that will complement a definition of autism that today is based purely on the observation of behaviors — a diagnostic method that, as we'll see in the next chapter, comes with its own perils.

4 Hiding and Seeking

YOU KNOW WHAT I HATE? The sound of hand dryers in public restrooms. Not so much when the air jet starts, but the moment someone's hands enter the stream. The sudden drop in register drives me nuts. It's like when the vacuum toilet on an airplane flushes. First comes the brief rainlike prelude, then a thunderclap of suction. I *hate* that. *Fingernails-on-a-chalkboard* hate.

You know what else I hate about air travel? The alarm that goes off when somebody in an airport accidentally opens a secure door. I hate alarms in general, for that matter. When I was a kid, the school bell made me absolutely crazy. It felt like a dentist's drill. No exaggeration: The *sound* caused a sensation inside my skull like the pain from a dentist's drill.

By now you've probably noticed a pattern in what I hate. I'm sensitive to sounds. Loud sounds. Sudden sounds. Worse yet, loud and sudden sounds I don't expect. Worst of all, loud and sudden sounds I *do* expect but cannot control — a common problem in people with autism. Balloons terrified me as a child, because I didn't know when they were going to pop.

Today I know that if I had been able to pop balloons myself, poking a small balloon with a pen and producing a soft sound, then working my way up to bigger and bigger balloons and louder and louder pops, I might have been able to tolerate balloons. I've heard a lot of people with autism say that if they can initiate the sound, they're more likely to be able to tolerate it. The same is true if they know the sound is coming; fireworks set off at random by kids down the block are shocking, but fireworks set off at the city park as part of a holiday program are acceptable. But when I was a kid, the same balloon that delighted and excited the other kids, the balloon that they wanted to toss to one another or flick with their fingers until it scraped the ceiling, I watched with dread. It loomed for me like a cloud of potential pain.

Our five senses are how each of us understands everything that isn't us. Sight, sound, smell, taste, and touch are the five ways — the *only* five ways — that the universe can communicate with us. In this way, our senses define reality for each of us. If your senses work normally, you can assume that your sensory reality is pretty similar to the sensory reality of everyone else whose senses are working normally. After all, our senses have evolved to capture a common reality — to allow us to receive and interpret, as reliably as possible, the information we need in order to survive.

But what if your senses don't work normally? I don't mean your eyeballs or eustachian tubes, the receptors on your tongue or in your nose or at the tips of your fingers. I mean your brain. What if you're receiving the same sensory information as everyone else, but your brain is interpreting it differently? Then your experience of the world around you will be radically different from everyone else's, maybe even painfully so. In that case, you would literally be living in an alternate reality — an alternate *sensory* reality.

I've been talking about sensory problems for as long as I've been giving lectures on autism, which is thirty years now. During that time, I've encountered people whose hearing fades in and out, so words go from sounding like a bad mobile phone connection to sounding like fireworks. I've talked to kids who hate to go into the gym because of the sound of the scoreboard buzzer. I've seen kids who can say only vowel sounds, possibly because they can't hear consonants. Almost all these people are autistic, and in fact, about nine out of ten people with autism suffer from one or more sensory disorders.

But pain and confusion don't affect just their lives. They also affect the lives of their loved ones. A normal child doesn't need to be told that the nonverbal autistic sibling requires more attention from their parents—that in many ways, the world of the family revolves around that child. For parents, taking care of even a normal child can be something of a full-time job; taking care of a child whose brain can't tolerate the motion of a parent crossing the room can be a full-*life* job. You can't take a child shopping or out to a restaurant or to the big brother's football game if the kid is going to be wailing in pain the whole time.

Besides, sensory disorders are not just an autism problem. Studies of nonautistic children have shown that more than half have a sensory symptom, that one in six has a sensory problem significant enough to affect his or her daily life, and that one in twenty should be formally diagnosed with sensory processing disorder, meaning that the sensory problems are chronic and disruptive. I myself have noticed that in a class I teach every semester, one or two of the sixty students have trouble drawing a cattle-handling system. They draw squiggly lines instead of smooth curves. I know they're not autistic, and they don't have astigmatism, but when I ask them what they see when they look at a page of print, they'll tell me that the letters are jiggling.

Yet what do we know about the science of sensory problems? Surprisingly little. It was surprising to me, anyway, once I started looking into the research on sensory problems.

For all the research on the autistic brain that neuroscientists and geneticists are conducting, for all the breakthroughs they're achieving, the subject of sensory problems is clearly not a priority. Sensory problems in people with autism are "ubiquitous," as a 2011 review article in *Pediatric Research* put it, yet the topic receives disproportionately little attention. Much of the research I found about sensory problems in autistics comes from nonautism journals, and many of those journals are not published in the United States. Even the articles on sensory problems in the autistic population that do appear in autism journals often go out of their way to bemoan the sorry state of research. "There is concern over the lack of systematical empirical research into sensory behaviors in ASD and confusion over the description and classification of sensory symptoms," wrote the authors of one 2009 study, while the authors of another study that same year complained of a "dearth of information." In 2011, I contributed an article to a big scholarly book on autism. More than fourteen hundred pages. Eighty-one articles in all. Guess what. The only paper that addressed sensory problems was mine.

Over the decades, I've seen hundreds if not thousands of research papers on whether autistics have theory of mind — the ability to imagine oneself looking at the world from someone else's point of view and have an appropriate emotional response. But I've seen far, far fewer studies on sensory problems — probably because they would require researchers to imagine themselves looking at the world through an autistic person's jumble of neuron misfires. You could say they lack theory of *brain*.

I suspect that they simply don't understand the urgency of the problem. They can't imagine a world where scratchy clothes make you feel as if you're on fire, or where a siren sounds "like someone is drilling a hole into my skull," as one autistic person described it. Most researchers can't imagine living a life in which every novel situation, threatening or not, is fueled by an adrenaline rush, as one study indicates is the case in many people with autism. Because most researchers are normal human beings, they're social creatures, so from their point of view, worrying about how to socialize autistics makes sense. Which it does, up to a point. But how can you socialize people who can't tolerate the environment where they're supposed to be social — who can't practice recognizing the emotional meanings of facial expressions in social settings because they can't go into a restaurant? Like other researchers, autism investigators want to solve the problems causing the most damage, but I don't think they appreciate just how much damage sensory sensitivity can cause.

I've talked to researchers who even say that the sensory problems aren't real. Hard to believe, I know. They call themselves strict behaviorists. I call them biology deniers. I tell them to consider this possibility: "Maybe that kid is freaking out in the middle of Walmart because he feels like he's inside a speaker at a rock concert. Wouldn't *you* be freaking out if you were inside a speaker at a rock concert?" I've had researchers then ask me, "If the kid is screaming because he's sensitive to sounds, then shouldn't *that* sound be bothering him?" Not if he's sensitive to only certain kinds of sounds. Sometimes those particular sounds don't even need to be loud in order to be annoying.

Not every person who suffers from a sensory disorder responds to a stimulus in the same way. I've seen children scream when a supermarket door opens swiftly, but I myself always found the movement of doors fascinating. One child will play with running water. Another will run away from a flushing toilet.

And not every person who suffers from a sensory disorder suffers to the same degree. I've learned to live with the sound of hands under air dryers or door alarms in airports. For some people, though, the sensory problems are debilitating. They can't function in normal environments like offices and restaurants. Pain or confusion defines their lives.

But whatever form these sensory problems take, they're real, they're common, and they require attention. I've given them that attention—and what I've found has surprised me, shocked me, and even led me to question some of the basic assumptions about autism itself.

While autism experts by and large have neglected sensory problems as a subject for research, the fact is you can't study autism without figuring out a way to categorize the sensory issues. I myself long ago accepted the traditional way of putting autistic people with sensory processing problems into three categories, or subtypes.

- Sensory seeking. This category covers problems that arise when the autistic person solicits sensations. Of course, we all seek sensations all the time. *What does that cake taste like? How will that linen shirt feel? Can I hear what the people sitting behind me on the bus are saying?* But autistic people with sensory problems tend to seek these sensations all the time. They can't get enough of them. They might crave loud noises or, in my case, deep pressure. They often stimulate these sensations through rocking, twirling, hand-flapping, or noisemaking.

The other two categories are sort of the opposite of the first category. Rather than seeking sensations, the people in these two categories are responding to unsolicited sensations.

- Sensory overresponsiveness. People with this are overly sensitive to input. They can't stand the smell of the pasta sauce, or they can't sit in a noisy restaurant or wear certain kinds of clothing or eat certain foods.
- Sensory underresponsiveness. People with this show poor or no response to common stimuli. For instance, they might not respond to their names, even though their hearing is okay, or they might not react to pain.

These three subtypes make a lot of sense. I never thought to question them. You see autistic people with sensory processing problems, and you can fit them into one category or another.

But some scientists have started rethinking these categories. In 2010, Alison Lane of Ohio State University as well as three collaborators published a paper titled "Sensory Processing Subtypes in Autism: Association with Adaptive Behaviors" in the *Journal of Autism Developmental Disorders*. (*Good,* I thought. *An article about sensory problems that's actually in an autism journal.*) As usual in papers about sensory processing, these authors were quick to point out how neglected their subject was: "Few studies have sought to investigate the relationship between SP [sensory processing] difficulties and the clinical manifestations of ASD." Then they got down to business.

The authors collected their data in the usual way. They relied on results from the Short Sensory Profile, a research tool that dates to the 1990s. Observers (usually parents) of people with sensory problems select which of thirty-eight behaviors match the behaviors of the subject. These behaviors correspond to seven sensory domains: tactile sensitivity; taste/smell sensitivity; movement sensitivity; underresponsive/seeks sensation; auditory filtering; low energy/weak; visual/auditory sensitivity. One indicator of tactile sensitivity, for instance, would be "Reacts emotionally or aggressively to touch." An item indicating movement sensitivity is "Fears falling or heights." Or under the heading of auditory filtering: "Is distracted or has trouble functioning if there is a lot of noise around."

After collecting the usual data, however, Lane and her collaborators subjected it to a different model of statistical analysis and discovered that sensory issues then fell into three slightly different categories. I don't need to go into the details of their methodology; you can look it up, if you're interested. Briefly, the new categories are:

- Sensory seeking, leading to inattentive or overfocused behavior.
- Sensory modulation (through either underresponsiveness or overresponsiveness) with movement sensitivity and low muscle tone.
- Sensory modulation (through either underresponsiveness or overresponsiveness) with extreme taste/smell sensitivity.

These categories, too, make a lot of sense at first. Extreme taste/smell sensitivity? I'd never thought of it as being separate from the other sensory problems, but sure, I could see the usefulness of framing a category that way. Low muscle tone? I've certainly met a lot of autistic people with floppy limbs and pasty skin. "[This] subgroup is particularly important to physical therapists," said a 2011 article in *Physical Therapy* that drew on Lane's research. "Children with ASDs who have atypical movement sensitivity usually are overresponsive to proprioceptive and vestibular input" — the sense of how the parts of the body work together and the sense of balance, respectively — "whereas children with low energy and weak motor responses have poor fine and gross motor skills."

Still, the idea that you could take the same data and create two different ways of organizing it — two different sets of categories — bothered me. Can both ways be valid? Can neither way be valid? What are these categories even telling us?

Then I realized: The problem isn't which way you interpret the data. The problem is the data itself.

Studies of severe sensory problems rely on the testimony of parents or caregivers. The conclusions in those studies rely on the methodology of the researchers. But why should we assume that all these interpretations reflect what's happening to the subjects themselves? A person who can't imagine living in a world of sensory overload is very possibly going to underestimate the severity of someone else's sensations and the impact on that person's life, and even misinterpret behavior as a sign of one sensory problem when it might be another.

If researchers want to know what it's like to be one of the many, many people who live in an alternate sensory reality, they're going to have to ask them.

Researchers routinely disparage self-reports, saying they're not open to scientific verification because they're subjective. But that's the point. Objective observation of behaviors can provide important information. But the person suffering from sensory overload is the only one who can tell us what it's really like. In my previous books, I've tried to describe my sensory problems, and other high-functioning autistics have also been able to describe the impact of sensory problems on their lives. But what about persons with far more severe, even debilitating sensory issues?

The problem in eliciting self-reports from this population is obvious. If a sensory problem totally disorganizes a person's way of thinking, then he'll have trouble describing the problem. If a person is nonverbal, then another means of expression, like typing or pointing, has to be used. In the most extreme cases, however, even that goal would be unrealistic. And unfortunately, wrist-supported writing produces unreliable information; the facilitator might be moving the hand without realizing it, as one would with the planchette on a Ouija board.

But overcoming the problems inherent in self-reporting is important. If the only self-reports about sensory issues that researchers have are from high-functioning adults, then the results are not representative. Sensory problems might be worse at lower levels of functioning; they might even be the *cause* of low levels of functioning. So a study that quotes only high-functioning autistics would present a wildly skewed view of the population. What's more, by adulthood, a person can develop coping mechanisms that disguise the true severity of the sensory problems and might not reflect the reality of the same problem as experienced by a frightened child.

I'm hoping that some of the new technologies might allow for a higher incidence of self-reporting. Tablets, for example, have a tremendous advantage over plain old computers, even laptops: You don't have to take your eyes off the screen. Usually typing is a two-step process. First you look at the keyboard, then you look at the screen to see what you've typed. That could be one step too many for someone with severe cognitive problems. Before tablets, a therapist would have to mount the keyboard of a desktop computer on a box so that it was right below where the print was appearing on the screen. In tablets, however, the keyboard is actually part of the screen, so eye movement from keyboard to the letter being typed is minimal. Cause and effect have a much clearer correlation. That difference could well be meaningful in terms of allowing people with extreme sensory problems to tell us what it's like to be them.

In the meantime, we have to rely on two self-reports from nonverbal individuals who can type. They're the only two who I can be sure are the authors of their words. I've examined both cases with an eye toward discovering what their sensory reality is like.

In his book *How Can I Talk If My Lips Don't Move? Inside My Autistic Mind,* Tito Rajarshi Mukhopadhyay describes his liberation from a locked-in autistic existence. It came in the form of a board filled with numbers and letters that his mother provided for him before he was four years old, in the early 1990s. With her help, he learned math and spelling. Eventually his mother tied a pen to his hand so that he could communicate through writing. Over the years Tito has published several books that describe how he experiences his reality in two parts: an "acting self" and a "thinking self." I recently looked back over his writing, and I recalled the first time I met him. And I understood that although I didn't realize it at the time, I had gotten to see both the acting self and the thinking self in very rapid succession.

I met Tito in a medical library in San Francisco. The lighting was low; if the library had any fluorescent lights, they had been turned off in anticipation of our visit. The room was silent, the atmosphere serene — free of distractions. The conversation involved just Tito, me, and his keyboard.

I showed him a painting of an astronaut riding a horse. I had deliberately chosen an image that he wouldn't have seen before — in this case, an advertisement for a technology company that I found in an old issue of *Scientific American* I'd grabbed off a nearby shelf. I wanted to see how he expressed himself in words. He studied the picture, and then he turned to his keyboard.

Apollo 11 on a horse, he typed rapidly.

Then he ran around the library flapping his arms.

When he returned to the keyboard, I showed him a picture of a cow.

We don't eat those in India, he typed.[1]

Then he ran around the library flapping his arms.

I asked him another question, but I no longer remember what it was. Still, you know what happened next. Tito answered it, and then he ran around the library flapping his arms.

And that was it for the conversation. Tito had done as much writing as he could in one session. He needed to rest, because even answering three short questions required tremendous effort.

What I had witnessed, I realize now, is Tito's acting self in action, the self that the outside world sees: a spinning, flailing, flapping boy. Which is also the self that Tito sees.

In his book, he described his acting self as "weird and full of actions." He saw himself as pieces, "as a hand or as a leg," and he said the reason he spun himself in circles was so that he could "assemble his parts to the whole." He recalled staring at himself in a mirror, trying to force his mouth to move. "All his image did was stare back," Tito wrote, adopting a third-person point of view that only underscored the disconnect between his acting self and his thinking self.

That self, his thinking self, is "filled with learnings and feelings." And frustrations. He recalled a doctor telling his parents that Tito couldn't understand what was happening around him, and he remembered his thinking self's unspoken response: "'I understand very well,' said the spirit in the boy."

The acting self runs around a library flapping his arms. The thinking self observes the acting self running around a library flapping his arms.

---

1. Note that Tito wasn't using the word *astronaut* or *cow.* He had to come in the back door, so to speak. He described the object rather than named it.

For me, the idea of two selves is reinforced by what Carly Fleischmann describes in her 2012 book *Carly's Voice: Breaking Through Autism,* which she wrote with her father, Arthur Fleischmann. For the first ten years of her life, Carly appeared to be a nonverbal autistic. Then one day she shocked her parents and her caregivers by suddenly working the keyboard on her voice output device. Prior to this eventful afternoon, Carly had used the device for one thing: she would touch the picture of an object or activity, and the electronic voice would speak the corresponding words. In fact, that very afternoon one of her therapists had been deleting items from the device in order to free up memory, and he had considered clearing the alphabet function. Fortunately, he didn't get around to it.

That day, when Carly arrived for her lessons, she was unusually restless and cranky and overall uncooperative. "What do you want?" one of the therapists asked her in exasperation, as if Carly were actually capable of answering. And she was! Carly grabbed the voice output device. "H-E-L-P T-E-E-T-H H-U-R-T," she typed, laboriously.

Carly was extremely low functioning. Like Tito, her acting self was in constant motion, sitting and rocking, screaming, trying to destroy everything within reach. Like Tito, her thinking self was taking in a lot more information than anyone would have thought. On some levels, her inner life was surprisingly normal. As Carly entered her teens, she developed what you might call typical teenage-girl interests. She had crushes on Justin Timberlake and Brad Pitt. When she appeared on a TV show, she found herself concentrating on a cute cameraman. But on other levels, her inner life was complicated in ways only she could know.

In one particularly striking scene in *Carly's Voice,* she invited her readers to imagine having a conversation in a coffee shop. If you're like most people, you would imagine yourself sitting across a table from someone who is talking to you, and you would imagine yourself listening closely.

Not Carly.

> For me that is a different case altogether. The woman who brushes along our table leaves an overpowering scent of perfume and my focus moves. Then the conversation over my left shoulder from the table behind us comes into play. The rough side on my left sleeve cuff rubs up and down on my body. That starts to get my attention, as the whoosh and whistle of the coffee maker blends into different sounds all around me. The visual of the door opening and shutting in the front of the store completely consumes me. I have lost the conversation, missing most of what the person in front of me is talking about. . . . I find myself only hearing the odd word.

At this point in the doomed conversation, Carly said, she would behave in one of two ways. Either she would shut down and become nonresponsive, or she would have a temper tantrum.

Now that's interesting, I thought when I read this passage. Imagine you're the person sitting across from her, and you have to describe her behavior for the sensory profile. If Carly shuts down—if she seems not to be listening to you, even though you're sitting directly in front of her, talking directly to her—you would categorize her as underresponsive. But if she throws a temper tantrum—if, as Carly said, she started "to laugh or cry or get mad or even scream for no reason you can pinpoint"—you would categorize her as overresponsive.

Two different behaviors, two different sensory-profile subtypes—at least, so it would seem if you were sitting across from her, watching her from the outside. But if you were Carly, living your life from the inside, the two reactions would have the same cause: sensory overload. *Too much information.*

Tito offered a similar scenario in his book. He described entering a room he had never entered before: He looks around, turning to different parts of the room, until he sees an object that intrigues him.

"The first thing I see is its color," he wrote. "If I do not get into a deeper cogitation of its color by defining it as 'yellow,' and mentally lining up all the yellow things I know of, including one of my yellow tennis balls when I was seven years old, I move to the shape" of the object. The object has a hinge, which he might or might not notice. But if he does notice it, then:

> I might get distracted by the functions of levers. However, I pull my attention from there and wonder about the function of that yellow, large rectangular object, with levers of the first order, called a hinge.
>
> Why is that yellow, large rectangular object with levers there? I mentally answer the question, "It has allowed me to come inside that room, and can be opened or closed. And what else can that be, other than a door." My labeling is complete.

And then he moves on to the next object in the room.

Tito also wrote about visiting a house and becoming lost in a magazine. He loved turning and touching "those smooth glossy pages," and he loved sniffing them too. Only afterward, when his mother discussed the visit and mentioned the pink roses on the lace curtains, and the piano, and a picture in a silver frame, did Tito realize that he had been so intent on the magazine, he'd missed everything else in the room.

From the outside, his behaviors in the two situations would seem different. Standing still, staring at the door, Tito would look as if he were underfocused, unengaged. Sniffing the magazine, he would look as if he were overfocused, too engaged. But, as with Carly in the coffee shop, even though the observable behaviors are different, the feelings behind them are the same.

These self-reports reinforce my longstanding hypothesis that some nonverbal autistics might be far more engaged in the world than they seem to be. They just happen to be living in such an extraordinary jumble of sensations that they have no way of productively experiencing the outside world, let alone expressing their relationship to it.

But these self-reports also demonstrate that Tito and Carly observe their own behaviors as closely as a parent or caregiver or researcher. Unlike those outside observers, however, they can tell us what their behaviors actually mean. The difference between the observer's view and the subject's experience — between the acting self and the thinking self — is the difference between what sensory problems *look like* and what they *feel like.*

I asked myself about my own experience with hearing difficulties as a child, when I would try to make sense out of the babble of adult voices talking too fast for me to follow. My hearing had two settings: Off, and Let All the Stimulation In. Sometimes I would shut down and block out all the stimuli. Sometimes I would throw a tantrum. Two behaviors, one feeling.

In the "Sensory Processing Subtypes" paper I mentioned earlier — the one that suggested a different way of organizing sensory problems — the authors noted that underresponsiveness and overresponsiveness "may co-exist" in the same child. Based on these examples, I would go further. If *responsiveness* refers to the visible response that a parent or caregiver or researcher observes, then fine — you can make a distinction. From an outside point of view, the child is underresponding or overresponding, underfocusing or overfocusing. The acting self exhibits two distinct types of behaviors. But if *responsiveness* refers to what the thinking self with the sensory problem is experiencing, then no — the distinction is meaningless. Underresponsiveness and overresponsiveness, or underfocusing and overfocusing, might be *the same thing.*

Does this possibility have any foundation in fact? I think it does.

I found anecdotal evidence in numerous descriptions in online self-reports that sounded similar to Carly's.

- "When lots of people are talking around me, at the same time, such as in a pub, I get overwhelmed and start to zone out, and can't make sense of any of it."
- "I just shut down and can't feel or react, so I usually just stand/sit absolutely still and stare very hard at something. Sometimes my mind is racing and that's very difficult to pull back."

- "I just need to sit quietly and refocus."
- "I often just become catatonic, with a stoic expression."
- "Your eyes try to go to every movement they perceive. That is part of what destroys your eye contact and makes you seem very inattentive."

What about scientific support? I found two papers hypothesizing that both underfocusing and overfocusing are caused by overstimulation. One paper, published in *Frontiers in Neuroscience* in 2007, proposed that autistics with sensory problems suffered from what the authors called "intense world syndrome." The authors wrote that "excessive neuronal processing may render the world painfully intense." To which the brain's response might be "to rapidly lock down the individual into a small repertoire of secure behavioral routines that are obsessively repeated." Another paper, published in *Neuroscience and Biobehavioral Reviews* in 2009, said that people with autism might be living in what the authors called "a world changing too fast." They can't follow what's happening around them, so they withdraw from their surroundings.

In either case, the lesson isn't that some people with autism receive too much information and are therefore overresponsive while other people with autism receive too little information and are therefore underresponsive. The lesson is that if your brain receives too much sensory information, your acting self might easily *look* underresponsive but your thinking self would *feel* overwhelmed.

The "World Changing Too Fast" paper offered several real-life examples from adults with autism, including one from me. I've postulated that the common autistic symptom of averting one's eyes "may be nothing more than intolerance for the movement of the other person's eyes." I've asked kids, "Why do you look out of the corner of your eyes?" They say, "Because I can see better that way." As for why they can see better that way, I don't know. Because the world is moving too fast and a sidelong glance makes all the motion less overwhelming? Maybe. I like that hypothesis, but without further research, that's all it is — a hypothesis.

I myself have been guilty of moving too fast for other autistic people. Daniel Tammet wrote that when he and I met, I quizzed him too quickly: "She spoke very fast, and I found it difficult to follow her." The autistic author Donna Williams wrote that "the constant change of most things never seemed to give me any chance to prepare myself for them." That's why, she said, she'd always loved the saying "Stop the world, I want to get off."

Or if not *stop* the world, at least slow it down. "The stress of trying to catch up and keep up," Williams wrote, "often became too much and I found myself trying to slow everything down and take some time out." One method she developed of slowing down the world was to rapidly blink her eyes or turn the lights on and off: "If you blinked really fast, people behaved like in old frame-by-frame movies, like the effect of strobe lights without the control being taken out of your hands." J.G.T. van Dalen, an adult with mild autism, was quoted in the "World Changing Too Fast" paper as saying he is "constrained to digest each object piece by piece." For him, this period of extraordinary focus doesn't feel normal. "Time seems to flow out rapidly," he said. For an observer, this period doesn't look normal either. The difference, he said, was that "a nonautistic person sees me as living slowly."

In each of these cases, the acting person would *look* slow to an observer. But the thinking person would *feel* the opposite.

The idea that hyperreactivity and hyporeactivity might be two sides of the same coin carries several important implications.

One is pharmacological. "While most [of] the commonly prescribed medication [tries] to increase neuronal and cognitive functioning, we conclude that the autistic brain needs to be calmed down," the "Intense World" authors wrote, "and cognitive functions need to be diminished in order to re-instate proper functionality." In my own experience, I found that when I began taking antidepressants to manage my anxiety—old-fashioned antidepressants like Zoloft and Prozac—the drugs calmed me down enough so I could learn social behaviors. And studies have shown that although risperidone (brand name Risperdal), an antipsychotic drug, doesn't directly affect the core deficit of social impairment, it does reduce the irritability that causes aggression. But I think it might also indirectly help overcome social impairment, because if you can manage the maladaptive behaviors, you at least have a chance to engage in the world in a more socially productive fashion.[2] (As always with prescription drugs, don't do anything without first consulting a doctor. And medication has to be dispensed very carefully; kids especially are sometimes accidentally overdosed.)

Another implication is educational. One of the common symptoms among persons with autism is a supposed inability to understand facial expressions. Yet a series of studies in the 1990s found that if children with ASDs watched facial expressions displayed slowly on video, they understood them equally as well as neurotypical children of the same developmental age. The "World Changing Too Fast" authors developed software that slowed down the presentation of visual and auditory cues. When ASD subjects were exposed to these gestures and sounds, they began imitating them, while normal subjects did not respond to the prompts because they'd long ago internalized these behaviors. Similarly, when these researchers slowed down spoken sentences, they found that ASD subjects experienced an increased understanding of meaning.

The idea that hyperreactivity and hyporeactivity are two variations on a theme might even have implications for theory of mind. The "Intense World" paper proposed that if the amygdala, which is associated with emotional responses, including fear, is affected by

---

2. For more on this topic, see chapter 6, "Believer in Biochemistry," in my book *Thinking in Pictures,* and chapter 7, "Medications and Biomedical Therapy," in my book *The Way I See It* (second edition).

sensory overload, then certain responses that look antisocial actually aren't. "Impaired social interactions and withdrawal may not be the result of a lack of compassion, incapability to put oneself into someone else's position or lack of emotionality, but quite to the contrary a result of an intensely if not painfully aversively perceived environment." Behavior that looks antisocial to an outsider might actually be an expression of fear.

Because dividing sensory problems into three subtypes now strikes me as an unreliable strategy, I'm going to do what I always do when I don't know enough about a topic. I ask myself, What *do* I know? And what I know about sensory problems is that we have five senses. So I'm going to arrange my discussion of sensory problems according to each one. (For ways to identify these symptoms and for practical tips to alleviate them, see the sidebar at the end of this chapter.)

Visual-Processing Problems

My visual processing is, if anything, superior to others', though I don't know whether that's due to how my eyes work or to how my brain interprets the signals that my eyes send. Nonetheless, I can say that at the age of sixty-five, I can still read a newspaper without glasses (though menus in dark restaurants and business cards with tiny print have started giving me some trouble). When I'm bored at a conference, I distract myself by looking at the fibers in the carpet. My night vision is so good that sometimes I forget to turn on the headlights.

Which is not to say I don't have some visual sensitivity. When I get tired, I'll start to see a halo around the light on a streetlamp or the flicker on an old TV-type computer screen. When I switch lanes on the highway, I have to make extra sure that I've left myself enough space. If a doctor asks me to hold my head still and follow a pen with my eyes, I hate it. Therapists tell me that my eyes jerk and I can't track smoothly.

At the other extreme are visual problems like the kind that author

Donna Williams, who is autistic, described in her writing: "Light refraction, i.e. shine, is a visual equivalent of noise reverberation and is a major source of visual overload. For someone sensitive to these things, the shine, or light refraction, can cause a visual effect of shooting out streams of sparks of light. This distracts from paying attention to other things, but this shine can also have the visual effect of cutting up people or objects." Thomas McKean, an autistic champion of self-advocacy, described this syndrome as Picasso vision, saying it was like "looking through broken glass or a cracked mirror."

On a more everyday level, I often encounter students with Irlen syndrome — named for Helen Irlen, an American therapist who found that certain writing and reading problems could be reduced or eliminated through the use of colored paper or lenses.[3] The idea is that white paper overwhelms a visual system that is sensitive to brightness, whereas the wavelengths of light in colored paper or lenses soothe it.

Having a mild case of Irlen syndrome — for instance, the print on a page jiggles a little when you're tired — isn't going to affect your academic performance. Colored lenses might help your eyestrain the same way that the reduced contrast on an e-book reader does. But I've seen severe cases where Irlen syndrome definitely interfered with a student's schoolwork — print was blurred, words moved, lines disappeared — and colored paper or lenses helped.

Sometimes I see students struggling with my design assignment. They may submit drawings that are full of wavy, squiggly lines instead of smooth arcs. I first suggest that they go to the counseling center, but sometimes, for whatever reason, they don't want to do that. So, fine. In that case, I send them to the copy shop, and I tell them to photocopy pages from a book using paper in all the differ-

3. It's also sometimes called Irlen-Meares syndrome; around the same time that Irlen was doing her research, a New Zealand teacher named Olive Meares described problems involving seeing black print on white paper.

ent pastel colors until they find the shade that helps them see better. It might be tan. It might be lavender. But one of the colors will work best.

I also send these students to the drugstore and tell them to try on sunglasses with lenses of various different colors, and the same principle applies there: You have to find the right color. "Don't buy what looks good," I tell them. "Buy what works." One day a student who had picked out pink-tinted lenses came rushing up to me. "Oh, Dr. Grandin," she said, all excited, "I got an A on my economics quiz!" Why? Because the PowerPoint slides stopped jiggling and she could finally read the numbers on the professor's graphs. As I always tell my students, it would be stupid to flunk out of school because you're not using tan paper or because you didn't make your computer background lavender!

It doesn't cost anything to try on sunglasses. You've got nothing to lose and everything to gain. I know a four-year-old girl who put on a pair of pink sunglasses that her parents had bought at Disneyland, and she went from being able to tolerate five minutes at Walmart to being able to handle an hour. It makes a big difference for parents if they can take a child shopping!

Auditory-Processing Problems

Over the years I've identified four auditory-processing problems as the most common.

- Language input. One type of language-input problem is not being able to hear hard-consonant sounds. When I was a child, I had difficulty differentiating hard-consonant sounds. To me, *cat, hat,* and *pat* sounded the same, because those consonants are quick. They're spoken fast. I had to figure out which was which by thinking about what word made sense in a particular context. This description certainly fits the "World Changing

Too Fast" article's hypothesis that I discussed earlier. The other type of language-input problem is hearing the words but not being able to connect meaning to them, a syndrome that Donna Williams calls being "meaning blind."

- Language output. I describe this problem as "a big stutter." As a child, I could understand the words that people spoke slowly but I could not get my speech out. The solution my speech therapist proposed was the same one suggested in the "World Changing Too Fast" paper: Slow down.
- Attention-shifting slowness. Once a sound has my attention, I have trouble letting go and moving on to the next sound. If a mobile phone rings during one of my talks, it totally disrupts my train of thought; it grabs my attention, and my ability to shift back is slower than most people's.
- Hypersensitivity to sound. The Internet is full of autistics' testimonials to the problem of loud and sudden sounds of all sorts — balloons, sirens, fireworks. But some of the problematic sounds are the kind that you would think would be more mundane: "I can't tolerate the sound of noodles being stirred (that horrid squishy sound)." Sometimes, though, hypersensitivity involves not a specific sound but a wealth of sounds: "You may have to ask the guy talking to you to repeat himself a few times because you were trying to get past the cars going by, the dog barking three blocks away, and the bug that buzzed past your ear."

Although these are the most common auditory problems that I've encountered, there are plenty more problems of a more specialized nature. For instance, I've seen a number of kids who are echolalic. These are the kids who can yak out TV commercials word for word. Their speech is just fine. But they have no idea what the words actually mean. Lots of times, they don't even understand that the meaning is in the words. They think it's in the tone of voice. Contrast this syndrome with my childhood problem of understanding what words meant but having trouble getting them out. I'm actually working on a brain-scan proposal to study these two types of syndromes side by side.

Whatever form they take, auditory problems seem to be especially prevalent in persons with autism. A 2003 study compared brain activation in response to speechlike sounds in five autistic and eight control subjects. The autistic subjects uniformly showed less activation in speech areas. Another study from 2003 compared the responses of fourteen autistic and ten control subjects to subtle changes in a sequence of repetitive sounds — what's called a mismatch field (MMF). Magnetoencephalographic (MEG) measurements of the controls uniformly indicated that their brains were detecting the changes, while MEG measurements of the autistic subjects' responses uniformly indicated that their brains were not.

And just to complicate matters, autistic people seem to get visual cues mixed up with aural cues. Normally when a person is listening, the visual cortex gets turned down. But a 2012 fMRI study found that when autistics were listening to sound cues, their visual cortices remained more active than neurotypicals'. If that's the case, then even while they're straining to process aural cues, they're being distracted and confused by visual cues.

But there's hope, and not just for autistics. Researchers have begun looking into the therapeutic effects of singing. Again and again I've heard from parents and teachers that they taught kids to talk through singing, and I wondered if there was a scientific basis for this relationship.

In healthy brains, the parts that appear to be related to language and music overlap to a high degree. Yet researchers have long noted that even nonverbal autistic patients exhibit a strong response to music. In a 2012 study conducted at the Columbia University Medical Center in New York under the supervision of Joy Hirsch (the same researcher we met in chapter 2), thirty-six nonverbal autistic subjects ranging in age from six to twenty-two were matched with twenty-one nonautistic controls ages four to eighteen. Through functional MRI, functional connectivity MRI, and DTI scans, the researchers found that during speech stimulation, activation in the left inferior frontal gyrus, which is closely associated with language, was reduced in the autistic sample relative to the controls. During song stimulation, however, the activation in the same area was greater in the autistic population than in the controls.

Until recently, though, very few studies on the use of music therapy in autistic subjects — never mind studies on the use of music therapy to help nonverbal autistics achieve speech — had been done. A 2005 study analyzed the data from forty autistic subjects, ranging in age from two to forty-nine, who had undergone two years of music therapy. All forty exhibited improvements in language and communication, as well as in behavioral, psychosocial, cognitive, musical, and perceptual/motor skills. And the parents or caregivers of all forty reported that the improvements extended beyond music and into other areas of the subjects' lives.

"Theoretically grounded music-based interventions have been underutilized, which is unfortunate because music perception and music making is known to be a relative strength of individuals with autism," the authors of a 2010 paper concluded. "In particular, no study has systematically investigated the efficacy of a music-based intervention in facilitating speech output, and whether an intensive program can induce plastic changes in the brains of these individuals. On the basis of previous and current research, we hope that such specialized treatments for autism will be developed in the near future."

One of the authors of that study — Catherine Y. Wan, of the Music and Neuroimaging Laboratory in the Department of Neurology at Harvard Medical School — not only hoped for specialized treatments but went on to help create one. It's called auditory-motor mapping training (AMMT), and it is designed to promote speech production directly by training subjects to experience the relationship between speaking at different pitches while tapping tuned electronic drum pads. "The therapist introduces the target words or phrases by simultaneously intoning the words and tapping the drums tuned to the same pitches," Wan wrote in a proof-of-concept study published in 2011. The paper reported that after having forty-five-minute individual sessions five times a week over an eight-week period, the six nonverbal children in the study, ages five to nine, showed "significant improvements in their ability to articulate words and phrases, with generalization to items that were not practiced during therapy sessions."

The paper concluded, poignantly if predictably, that the number of interventions of this type currently in use are "extremely limited." So do we have conclusive scientific proof that music therapy facilitates communication in nonverbal autistic children? No. But I'd bet that the anecdotal evidence I've heard over the years, from teachers and parents, is right.

Touch and Tactile Sensitivity

As the person who felt the need to invent the squeeze machine to counter her anxiety and panic attacks, I obviously have a strong case of touch sensitivity — and I've written extensively about it in my other books. But my tactile problems don't stop there. Clothing drives me crazy if it's not the right texture. I've gotten a lot of T-shirts as presents from members of the audience at my public talks. Sometimes the T-shirts are scratchy, and sometimes they aren't, even if they're all made of 100 percent cotton and I've washed them to soften the fabric. The difference, believe it or not, is something in the weave or in the type of cotton.

What other tactile experiences present problems? You'd be surprised. Here are some examples from the website Wrong Planet (wrongplanet.net) about autistic sensitivities involving the sense of touch.

- "I simply cannot stand wet sand. Enforced beach holidays were hell for me."
- "I am utterly incapable of touching soft things . . . teddy bears, very soft blankets, etc., especially when my hands are dry. The thought of that disgusts me beyond words." (This person's solution would drive me absolutely crazy: "roughest, lowest thread count sheets I can find.")
- "Wet sand, cream, and towels. These can be combined into my worst possible combination, which is sun-lotioned skin, covered in sand, wiped off with a wet towel."
- "Wet sleeves."
- "I can't stand the feel of newsprint—it feels like teeny-tiny splinters all in my fingertips."
- "Sponges are pretty nasty, although oddly enough I did used to quite enjoy eating sponge."
- "Every time I wear something that is not loose fitting enough my skin feels like there are small insects crawling all over it."
- "I HATE HATE HATE the feeling and texture of denim jeans. It's so dry and scratchy."
- "Petting a dry dog with wet hands."
- "Glass that has come out of a dishwasher—it feels squeaky in an awful way."

Olfactory Sensitivity and Abnormal Smell / Taste Sensitivity

Some people just cannot tolerate certain smells. They walk down the detergent aisle in the supermarket, and they're overwhelmed. My collaborator Richard has a friend who gets headaches from the smell of newsprint. Growing up, she used to dread the fat Sunday edition. Today she reads newspapers only online.

Some people just cannot tolerate certain tastes. A lot of times, this aversion has to do with texture. I don't like slimy things. Runny egg whites? *Yuck.* (Although what seems like a taste sensitivity might actually be an auditory problem. For some people, the crackle of a potato chip heard inside the cranium is unbearable.)

As with tactile sensitivity, the range of triggers is astonishing:

- "Any grain or carb that is soggy."
- "Flat soda — once it is open more than a minute I won't drink it."
- "Taco seasoning makes me dizzy."
- "I never ate at a fish restaurant in my life. Just driving by one makes me gag. I can't stand the smell of it."

Researchers might not trust self-reports, but to my mind these quotations are an invaluable resource, not just for the information they contain but for the larger lesson: If you want to know what the symptoms of autism mean, you have to go beyond the behavior of the autistic person and into his or her brain.

But wait. Isn't the diagnosis of autism based on behaviors? Isn't our whole approach to autism a result of what the experience looks like from the outside (the acting self) rather than what the experience feels like from the inside (the thinking self)?

Yes. Which is why I believe the time has come to rethink the autistic brain.

Visual-Processing Problems

How to Identify a Person with Visual-Processing Problems:

- Flicks fingers near the eyes
- Tilts head when reading, or looks out of the corner of the eye
- Avoids fluorescent lights (This problem is especially prevalent with fluorescent lights that operate at 50 to 60 Hz cycles.)
- Fears escalators; has difficulty figuring out how to get on and off

- Acts blind while negotiating an unfamiliar setting, such as a stairway in a strange house
- Sees wiggling print when reading
- Has poor night perception; often hates driving at night
- Dislikes rapid movement; avoids automatic sliding doors and other things that move rapidly (and/or unexpectedly)
- Dislikes high contrast of light and dark; avoids bright contrasting colors
- Dislikes multicolored floor tiles and anything that forms a lattice or grid

Practical Tips for People with Visual-Processing Problems:

- If you're under fluorescent lights, wear a hat with a brim, sit next to a window, or bring your own lamps with old-fashioned incandescent bulbs.
- Get Irlen lenses or experiment with different pale-colored sunglass lenses.
- Print reading materials on tan, light blue, gray, light green, or other pastel paper to reduce contrast, or use transparent colored overlays.
- Get a laptop or a tablet rather than one of the older desktop computers with flickering screens. Try colored backgrounds.

Auditory-Processing Problems

How to Identify a Person with Auditory-Processing Problems:

- Sometimes appears deaf even though auditory threshold is normal or near
- Can't hear when there's background noise
- Has difficulty hearing hard consonants; hears vowels more easily
- Covers ears around loud sounds

- Has frequent tantrums in noisy places such as train stations, sports arenas, loud movie theaters
- Ears hurt from certain sounds such as smoke alarms, firecrackers, balloons popping, and fire alarms
- Hearing shuts off or changes volume, especially when in overstimulating environments; noises may sound like bad mobile-phone connections
- Has difficulty localizing the source of a sound

Practical Tips for People with Auditory-Processing Problems:

- Wear earplugs in noisy places (but take them out for at least half the day, to prevent hearing from becoming more sensitive).
- Record sounds that hurt the ears on a recording device, then play them back at a reduced volume.
- Loud sounds and noises are more easily tolerated when you are rested and not tired.
- Loud sounds are better tolerated when you either initiate them or know they're coming.

Touch and Tactile Sensitivity

How to Identify a Person with Tactile Sensitivity:

- Pulls away when hugged by familiar figure
- Takes off all clothes or wears only certain articles (wool and other scratchy materials cause the most problems)
- Can't tolerate certain fabrics or textures
- Seeks deep-pressure stimulation by getting under heavy pillows or carpets, rolling up in blankets, or squeezing into tight spots (for instance, between a mattress and box spring)
- Lashes out or throws a tantrum when lightly touched

Practical Tips for People with Tactile Sensitivity:

- Deep pressure can help desensitize an individual; it can also help teach feelings of kindness. Most individuals with autism can be desensitized and can learn to tolerate being hugged by, for instance, wearing weighted vests, getting under heavy cushions, or receiving firm massages.

- Sensitivity to scratchy clothing is more difficult to desensitize, but try washing all new clothing several times before letting it touch the skin; remove all tags; and wear underwear inside out (which gets the seams away from the skin).
- Sensitivity to medical examinations can sometimes be desensitized by applying deep pressure to the area that has to be examined.

Olfactory and Taste Sensitivity

How to Identify a Person with Olfactory Sensitivity:

- Avoids certain substances and smells
- Is attracted to certain strong smells
- Throws a tantrum in the presence of some smells

How to Identify a Person with Taste Sensitivity:

- Eats only certain foods
- May avoid foods with certain textures

Practical Tips for People with Olfactory/Taste Sensitivity:

There's an old vaudeville joke: Man walks into a doctor's office, raises his hand over his head, and says, "Doctor, it hurts when I do this!" To which the doctor says, "Well, don't do that!"

That's pretty much what I have to say about these two categories. If you don't like it, well, don't do it! If the smell the person is attracted to is something nasty, such as feces, try substituting a source of a strong, pleasant smell, such as peppermint, or other odors used for aromatherapy.

Part II

RETHINKING THE AUTISTIC BRAIN

5 Looking Past the Labels

I HAD MY EYE on Jack. He was ten years old, and he had taken only three skiing lessons in his life. I was in high school, and I'd been taking skiing lessons for three years. Yet I would watch Jack pass me on the slope, and I would see him execute these gorgeous stem christie turns, and, man, he could handle the four-foot ski jump with no problem. Meanwhile, I was still working my way up to *one* good christie, and every single time I tried the ski jump, I fell, until I was scared to use it.

What was so special about Jack?

Nothing, it turns out. What was so special, instead, was me — me and my autism. The connection between my autism and my poor athletic performance is pretty obvious in retrospect. At the time, though, I didn't see it. Not until I was in my forties and I had the brain scan showing that my cerebellum — the part of the brain that helps control motor coordination — is 20 percent smaller than normal did I put two and two together. Now it all made sense! I couldn't keep my skis together without falling because —

Because what? Because I'm autistic? Or because I have a small cerebellum?

Both answers are correct. Which, however, is more useful? That depends on what you want to know. If you're looking for a label, something that will help you understand who I am in a general sense, then "because I'm autistic" is probably good enough. But if you're looking for how I got that way specifically—if you're looking for the biological source of the symptom—then the better answer is definitely "because I have a small cerebellum."

The difference is important. It's the difference between a diagnosis and a cause.

My research on subtypes of sensory problems for the previous chapter got me thinking about the limitations of labels. I realized that two different labels—underresponsive to sensory input and overresponsive to sensory input—can describe the same experience: *too much information!* The labels might be useful, but, as in the skiing example, their usefulness depends on what you want to know. Do you want to know what the behavior looks like from the outside? Or do you want to know what the experience feels like from the inside? Do you want a description for a set of symptoms—a diagnosis? Or do you want a source for a particular symptom—a cause?

Parents come up to me all the time and say things like, "First my kid was diagnosed with high-end autism. *Then* he was diagnosed with ADHD. *Then* he was diagnosed with Asperger's. What is he?" I understand their frustration. They're at the mercy of a medical system that's full of label-locked thinkers. But the parents are part of that system too. They'll ask me, "What's the single most important thing to do for an autistic kid?" Or "What do I do about a kid who misbehaves?" What does that even *mean?*

I call this kind of thinking label-locked because people get so invested in what the word for the thing is that they no longer see the thing itself. I've encountered the same kind of label-locked thinking elsewhere as well. A livestock handler might say to me, "My horse is wild. What should I do?" Or someone who has read my books on animal behavior will say to me, "My dog's crazy. What should I do?" Well, first you have to tell me what *wild* or *crazy* even means in each case. I don't have a clue unless you give me one. Does the dog try to bite the hands of strangers? Or does he jump on people because he's really happy?

What I say in all these cases is the same: Don't worry about the label. Tell me what the problem is. Let's talk about the specific symptoms.

First question I ask parents who want me to advise them is "How old is the kid?" What I might recommend for a three-year-old is going to be completely different from what I might recommend for a sixteen-year-old.

Next question is "Does the kid talk?" If he's nonverbal, that's one thing. Let's start trying to teach him and see what happens. If he's verbal, I'll say, "How good is his speech?" If the description is too vague, I'll say, "Give me an example." I want to know if the child is speaking in complete and grammatically correct sentences. Does he speak only in single words? Does he pronounce words accurately or does he say, as I did, *buh* for *ball*?

Can the kid hold a conversation? Can he place an order at a fast-food counter? If not, then the first thing you want to do is teach the kid social skills, starting with taking turns and saying "Please" and "Thank you."

Does she have trouble making friends? Is she in school? Does she have a favorite subject?

The questions can go on and on, of course, just as they can for anybody—autistic or not. We're all individuals. We all have a range of skills, habits, preferences, limitations. What would a totally normal brain even be like? A brain that is average in every way, that has the average number of neural connections, the average size of amygdalae and cerebellum, the average length of corpus callosum?

It would probably be pretty boring.

The differences are what makes us individuals—the departures from the norm, the variations in the brain. Take the corpus callosum, which is the collection of neural cables that stretch the length of the brain and connect the left and right hemispheres. I have more of those cables than normal, but obviously someone can have even more than I do, or fewer than I do, or the normal amount, or fewer than normal. And my brain's language circuit branches more than a normal brain's, but, again, the extent to which language circuits branch exists on a continuum. The cerebellum size that probably affects my skiing — another continuum. The number of de novo copy number variations in someone's DNA? The particular position of those CNVs on the chromosome? Continuum and continuum. I have often thought that eventually we're going to be asking ourselves at what point this or that autism-related genetic variation is just a normal variation. Everything in the brain, everything in genetics — they're all one big continuum.

The addition of Asperger's to the *DSM-IV* in 1994 validated the idea of an autistic spectrum, but the meaning of "on the spectrum" itself has changed over the years. "In scientific circles," a 2011 article in *Nature* reported, "many accept that certain autistic traits — social difficulties, narrow interests, problems with communication — form a continuum across the general population with autism at one extreme."

In other words, you don't have to have an autism spectrum disorder diagnosis to be "on the spectrum."

This notion was popularized by the psychologist Simon Baron-Cohen. In 2001, he and his colleagues at the Autism Research Centre in Cambridge, England, introduced the autism-spectrum quotient questionnaire. People often take the AQ test online just to see whether they fall on the autistic spectrum. They might be wondering if they have Asperger's or high-functioning autism. Or they might want to see what traits they have that, if amplified, would qualify them for one of those labels.

If nothing else, the AQ test got a lot of people thinking about behavior in a new way — the behavior of autistics, certainly, but the behavior of nonautistics too. Their own behavior. The behavior of a neighbor, or a coworker, or oddball Uncle Ned with his disturbingly thorough stamp collection. Behavior that previously had seemed peculiar or perhaps aggressively strange now made a kind of sense.

The test consists of fifty statements. (See appendix.) For each statement, you choose from four responses, ranging from "definitely agree" to "definitely disagree." Definitely agreeing with the statement "I would rather go to a library than a party" might indicate that a person has an autistic bent. Definitely agreeing with the statement "I find myself drawn more strongly to people than to things" would suggest a more neurotypical person. When Baron-Cohen and his colleagues administered the test in a clinical setting, the average score in the control group was 16.4 out of 50, while 80 percent of those diagnosed with autism or a related disorder scored 32 or higher. But if you scored 33, would you be autistic? Not necessarily. What about 36? Or 39? What is the cutoff point?

Label-locked thinkers want answers.

This kind of thinking can do a lot of damage. For some people, a label can become the thing that defines them. It can easily lead to what I call a handicapped mentality. When a person gets a diagnosis of Asperger's, for instance, he might start to think, *What's the point?* or *I'll never hold down a job.* His whole life starts to revolve around what he *can't* do instead of what he *can* do, or at least what he can try to improve.

Label-locked thinking goes the other way too. *You* might be comfortable with your diagnosis but worry that it will define you in the eyes of others. What will your boss think? Your coworkers? Your loved ones? Half the employees at Silicon Valley tech companies would be diagnosed with Asperger's if they allowed themselves to be diagnosed, which they avoid like the proverbial plague. I've been to their offices; I've seen the work force up close. Many of the hits on my home page come from Silicon Valley and other areas with a high concentration of tech industries. A generation ago, a lot of these people would have been seen simply as gifted. Now that there's a diagnosis, however, they'll do anything to avoid being ghettoized.

Label-locked thinking can affect treatment. For instance, I heard a doctor say about a kid with gastrointestinal issues, "Oh, he has autism. That's the problem" — and then he didn't treat the GI problem. That's absurd. Just because gastrointestinal problems are common in people with autism doesn't mean that the GI problems are untreatable on their own. If you want to help the kid with GI issues, talk about his diet, not his autism.

And label-locked thinking can affect research. "One of the curses in this field," a study on vision in autism concluded, "is the size of the error bars, which always seem to be at least twice as large in the ASD data compared to the controls." Error bars twice as large as the controls' error bars? Right there, that should tell you that you have a huge variation in the sample — that you have subgroups in the population that need to be identified and separated out. You throw people with Irlen syndrome and people who look out of the sides of their eyes into the same sample and you'll end up comparing apples and oranges. The error bars aren't a *curse.* They're an obstacle that the researchers have created for themselves and then placed in their own path.

The same is true for studies that conclude that some solutions to sensory problems, like weighted vests or Irlen glasses, don't work for people with autism. I used to read these studies, and I would tell myself, *But I've* seen *weighted vests work, again and again!* The problem with the research, I've realized, is that autistic people don't all have the same sensory problems. If you have twenty people with autism, shaded glasses or weighted vests will help maybe three or four. So researchers say, "Well, look — these devices work on only 15 or 20 percent of the autistic population!" So what? That result doesn't mean that colored glasses don't work for autism; it means that colored glasses *do* work for autistics with certain specific visual problems. They work on a *subgroup* of the autistic population.

I'm not saying that we shouldn't use labels. Of course we should. Without the label that Leo Kanner gave it, autism might have gone undiagnosed, untreated, and just plain ignored. Labels have been incredibly important, and they will continue to be incredibly important. For the purposes of medicine, educational benefits, insurance reimbursements, social programs, and so on, they're necessary. And if you're a researcher looking into autism, sometimes it makes sense to test only autistic subjects against controls.

But sometimes it doesn't, because autism is not a one-size-fits-all diagnosis.

However the APA defines autism, the diagnosis is going to be imprecise. That's the nature of a spectrum. The lack of a diagnosis was what the first formal set of standards in the *DSM-III* tried to correct, and the lack of precision in the diagnoses for autism and autism-related disorders was what subsequent editions tried to correct. Unfortunately, I don't think the latest effort — the *DSM-5* — is going to be much help in clearing up the confusion, and in some ways, it's only going to complicate the situation.

In the *DSM-IV*, a diagnosis of autism depended on three criteria, called the triad model. Those criteria were:

- Impairment in social interaction.
- Impairment in social communication.
- Restricted, repetitive, and stereotyped patterns of behavior, interests, and activities.

The first two might sound similar to each other in that they both involve issues of socializing. In fact, that's the official justification for collapsing them into one criterion for the *DSM-5*. In a 2010 presentation before the federal Interagency Autism Coordinating Committee, the chair of the *DSM-5* Neurodevelopmental Workgroup said, "Deficits in communication are intimately related to social deficits. The two are 'manifestations' of a single set of symptoms that are often present in differing contexts." As a result, the *DSM-5* uses a two-criteria, or dyad, model:

- Persistent deficits in social communication and social interaction.
- Restricted, repetitive patterns of behavior, interests, or activities.

I understand why the APA might consider changing from a triad to a dyad model. The idea of separating the social from the behavioral does have a basis in science; the two domains are in fact biologically different. In lab tests on mice, researchers have shown that risperidone, an antipsychotic drug, does not affect social behaviors but does affect fixated behaviors — possibly because it sedates the mice. Conversely, researchers have shown that the social behavior of mice was improved by training, but the fixated behavior was not. Those results alone tell us that repetitive behaviors and social problems operate in separate systems in the brain. So a dyad system that recognizes the distinction between those two systems does make sense.

What isn't scientific about the *DSM-5*'s handling of the diagnostic criteria, however, is its collapsing together social interaction and social communication. Social interaction covers nonverbal *behavior* that involves being with another person — making eye contact, smiling, and so on. Social communication covers the verbal or nonverbal *ability to converse* — share ideas and interests, for example. Do impairments in social communication and impairments in social interaction actually belong to one single domain? Does an inability to get words out and master grammar and syntax (known as specific language impairment or syntactic-semantic disorder) really come from the same place in the brain as a tendency to speak with abnormal intonation and give conversational responses that are socially inappropriate (known as pragmatic language impairment or semantic-pragmatic disorder)? Are language mechanics and social awareness closely related, neurologically speaking? I doubt it — and I'm not alone in that doubt.

A 2011 paper in the *Journal of Autism and Developmental Disorders* surveyed more than two hundred fMRI and DTI studies in an effort to determine whether the dyad model has a basis in neuroimaging data. The authors' conclusion: "only partially." They found that neuroimaging supports the separation of behavior and communication into two categories. No surprise there. But they also found that neuroimaging supports the division of communication into two further categories, just like the *DSM-IV* said — though not necessarily the two categories the *DSM-IV* described!

The *DSM-5* is also changing the scope of the diagnosis itself. In the *DSM-IV,* the autism-related category was pervasive developmental disorders, and it included[1] these diagnoses:

- Autistic disorder (also called "classic" autism)
- Asperger syndrome
- Pervasive developmental disorder not otherwise specified (or atypical autism)

The *DSM-5* lists one:

- Autism spectrum disorder

So, you might ask, what happened to Asperger's and PDD-NOS? Let's take them one at a time.

The big change regarding Asperger's and autism is speech delay.

1. It also included Rett syndrome and childhood disintegrative disorder, which don't concern this discussion.

Previously, if you had speech delay as a kid, as I did, then you fell on the autistic side of the diagnostic divide (assuming you met the other necessary criteria, of course). If you didn't, then you fell on the Asperger's side. Now some of the former Aspies will get an ASD diagnosis, just by virtue of meeting all the criteria for that diagnosis but not having speech delay.

The APA says that those already diagnosed with autism will keep the diagnosis. But what about the previously undiagnosed Aspies who meet only the social half of the new dyad criteria — deficits in social communication and interaction but not in repetitive behaviors and fixated interests? They could find themselves in another subcategory altogether: communication disorder. Specifically, they'll find themselves receiving a diagnosis that's new to the *DSM*: social communication disorder. Which is, basically, autism without the repetitive behaviors and fixated interests. Which is, basically, rubbish. (To my way of thinking, social impairments are the very core of autism — more so than the repetitive behaviors.) So having a diagnosis of social impairment that's distinct from the diagnosis of autism is the same as having a diagnosis of autism that's distinct from the diagnosis of autism!

Those who previously would have been diagnosed with Asperger's might learn that they don't belong in the neurodevelopmental-disorders category at all, at least not officially. They could find themselves in a whole other diagnostic category: disruptive, impulse-control, and conduct disorders. The decision ultimately comes down to an individual doctor's opinion — and if you say that that doesn't sound like science, I wouldn't disagree.

First, as a biologist, I find just about this whole diagnostic category scientifically suspect. The category includes six diagnoses. As far as I can see, only one has any basis in science: intermittent explosive disorder. Neuroimaging shows that if you lack top-down control from the frontal cortex to the amygdala, you'll be prone to outbursts that will get you fired or arrested. But as for the other diagnoses in the disruptive, impulse-control, and conduct disorders category? I smell a strong case of "If we label them that, then we don't have to give them ASD services and we can just let the police deal with them." The *DSM* might as well call this category Throw 'Em in Jail.

Second, these diagnoses overlook the gifted but frustrated—the typical Aspie or high-functioning autistic who is laboring in a non-sympathetic environment. Consider the oppositional defiant disorder diagnosis: "The disturbance in behavior causes clinically significant impairment in social, educational, or vocational activities." I guarantee you that if you take a third-grader who can read high-school math texts and make him do baby-math drills over and over and over, he will turn oppositional defiant—because he's bored out of his mind.

How do I know? Because I've seen these cases—kids who are considered to have severe behavior problems at school until you give them math lessons that meet them where their brains are. Then their behavior normalizes, and they become productive and engaged—maybe even model students.

And here, again, we see the limitations, and even dangers, of label-locked thinking—the difference between what behavior *looks* like from the outside and what it *feels* like from the inside.

As for PDD-NOS, *DSM-IV* used this catchall diagnosis to describe several scenarios, including atypical autism, defined as "presentations that do not meet the criteria for autistic disorder because of late age of onset, atypical symptomatology, or subthreshold symptomatology, or all of these." In the *DSM-5,* though, people with that diagnosis might find themselves jettisoned from autism altogether and put into another neurodevelopmental-disorder subcategory, intellectual-development disorders—specifically, intellectual or global developmental delay not elsewhere classified. No wonder so many parents feel like they're in the Diagnosis of the Year club.

For a lot of people, the changes to the *DSM* won't make a difference. For instance, under the *DSM-5* guidelines, I would be diagnosed with autistic spectrum disorder. If you look at the description of what constitutes social impairments and repetitive behaviors, I definitely qualify. Extreme distress at small changes? That was me as a kid. Fixated interests? Boy, I had that. Hypersensitivity to sensory input? Let me tell you about the squeeze machine.

But for a lot of people, these changes will make a huge difference. A 2012 survey of 657 people who had been clinically diagnosed with any one of the three *DSM-IV* autism spectrum disorders found that 60 percent would continue to receive the ASD diagnosis under *DSM-5* criteria but 40 percent would not. Breaking those numbers down into subgroup diagnoses, the researchers discovered that 75 percent of subjects who had received the specific diagnosis of autism according to *DSM-IV* criteria would also meet the *DSM-5* criteria for ASD, but only 28 percent of those diagnosed with Asperger's would, and only 25 percent of those diagnosed with PDD-NOS would.

A later study that concentrated on only the PDD-NOS diagnosis reached a far more optimistic conclusion: nine out of ten children with a *DSM-IV* PDD-NOS diagnosis would be eligible for a *DSM-5* ASD diagnosis. The disparity between the two reports, however, should give any parent, let alone scientist, pause.

What practical effects will these diagnostic changes have? Will people who were labeled Asperger's and are now labeled autistics experience a different response from the world? From themselves? How will these changes affect insurance coverage? What about social services? Autistics have more problems than those with Asperger's; will they still get the same range of help as before? That question will be decided on a state-by-state basis, but these changes have opened a Pandora's box of possibilities.

And research! Any study of autism that uses *DSM-5* criteria for autism is going to be mixing speech-delay apples and non-speech-delay oranges. For instance, we've seen in the literature that sensory problems tend to be a whole lot worse among members of the population who have speech delays. How are researchers going to be able to compare *DSM-5* sensory-problem studies with pre-*DSM-5* studies?

To me, the *DSM-5* sounds like diagnosis by committee. It's a bunch of doctors sitting around a conference table arguing about insurance codes. Thanks to label-locked thinking, we now have a cornucopia of diagnoses — and there simply aren't enough brain systems for all these names.

Back in 1980, when the *DSM-III* first tried to codify the diagnosis of autism, nobody knew about brain systems. Nobody knew much about DNA sequencing. But now we do. We may not be able to apply those advances in science to the *DSM* yet, but what we can do, I feel, is begin to change the way we think about the autistic brain. Instead of talking about *sets of symptoms* in an attempt to assign them a label, we can begin to talk about *one particular symptom* and attempt to determine its source. We've reached a point in our research that we can match symptoms and biology.

For the first thirty years or so after Leo Kanner introduced the term *autism*, in 1943, the emphasis in the psychiatric community was on finding a cause, and because psychoanalytic theory dominated the psychiatric thought of the day, the cause was hypothesized to be the behavior of the parents, especially the mother.

Let's call this period phase one in the history of autism, and let's say it extended from 1943 to 1980, the year the American Psychiatric Association published the *DSM-III*.

That edition of the *DSM* represented a shift in the psychiatric community toward greater scientific rigor in its treatment of mental illnesses, a shift that included the first official diagnosis for autism. Since then, much of the discussion about autism has involved what specific symptoms make up the diagnosis.

Let's call this period phase two in the history of autism, and let's say it extended from 1980 to 2013, the publication year for the *DSM-5*.

The diagnosis can and will continue to change, but now we can shift our emphasis once again. Thanks to advances in neuroscience and genetics, we can begin phase three in the history of autism, an era that returns to the phase-one search for a cause, but this time with three big differences.

One, the search for the cause involves not the mind but the *brain* — not some phantom refrigerator mom but observable neurological and genetic evidence.

Two, because we realize how extraordinarily complex the brain is, we know this search will lead not to *a* cause but to *causes.*

Three, we need to be looking for a cause or multiple causes not of autism but of *each symptom* along the whole spectrum.

Phase-two thinking says, *Maybe I can't ski well because I'm autistic.* Phase-three thinking says, *Maybe I can't ski well because I have a smaller than normal cerebellum.*

Phase-two thinking says, *Let's group people together by diagnosis.* Phase-three thinking says, *Forget about the diagnosis. Forget about labels. Focus on the symptom.*

Instead of — or at least in addition to — assigning human subjects to studies by their autism diagnosis, we should be assigning them by their main symptoms. As I learned from examples like Carly Fleischmann's description of feeling overstimulated in the coffee shop, I think researchers should stop pooh-poohing self-reports and start looking at them very carefully and, in addition, begin eliciting them from subjects in new ways. Then they should be putting the subjects into studies based on those self-reports.

I once had a graduate student who saw wavy lines between the curved lines in a drawing of a cattle chute, and sometimes she saw only pieces of words. She wasn't autistic, but these symptoms were notably similar to those described by Donna Williams, who definitely is autistic.

I say, Throw 'em both in a scanner, and let's see what lights up. Let's see where the problem is. Is it in the language-output area? Language-meaning?

Let's take the people who can't ride on an escalator because they can't figure out how to get on and off. Or let's take the people who hate driving at night. Let's take those subgroups and put them against controls who don't have that problem. Let's take this secretary over here who can type 180 words a minute. Let's take another secretary who can type 90 words a minute. Let's throw them both in a scanner and compare them, motor cortex to motor cortex.

Some researchers, I'm pleased to see, are beginning to recognize the limitations of labels. And they're beginning to recognize the need for narrower definitions of targets. A 2010 article, "Neuroimaging of Autism," concluded: "For autism it becomes more and more clear that the possibility to identify one single marker might become very small, just because of the large variability we meet in [this] spectrum. In this view the definition of *smaller autism subgroups* with *very specific characteristics* might give a key to further elucidate this complex disease" (emphases added).

Personally, I would go even further and argue that we need to think not just about smaller autism subgroups that are defined by their symptoms but about the symptoms themselves. Because thinking about individual symptoms on a symptom-by-symptom basis will eventually allow us to think about diagnosis and treatment on a patient-by-patient basis.

My friend Walter Schneider, who developed high-definition fiber tracking at the University of Pittsburgh, is already making that argument — probably because he has so vividly seen for himself the potential of this approach.

"We are searching for actionable diagnosis," he says. "Not just that we say, 'Yeah, you're different,' but, 'You are different and because of this particular form of difference, we think this is the most likely path for getting you to as much of the outcome as we want you to get.' We want to go in and in on that individual brain — not a group study but an individual brain — so we can say to a parent, 'This is what the situation is, this is what we expect the effect to be, and this is how we plan to get around it as efficiently as possible to give you effective communication with your child in the next two years.'"

You can hear the same argument beginning to surface in genetics as well. Yale neurogeneticist Matthew W. State likes to invoke the medical phrase *from the bench to the bedside* — meaning from experiments on groups to treatments for individuals. In a 2012 article in *Science,* he and collaborator Nenad Šestan suggested that researchers look for inspiration from other areas of medicine that have made this transition. "For example, heart disease and stroke prevention both rely in part on the management of hypertension," they wrote. "It may well be that ASD and schizophrenia will increasingly be thought of in a similar light" — different behaviors arising from the same genetic source. As a result, Šestan and State anticipated that treatment trials would be organized around "shared mechanisms" rather than "psychiatric diagnostic categories." They didn't doubt that this rethinking of the autistic brain would be challenging. But like Schneider, they foresaw the development of therapies that were not only more effective but "more personalized."

Twenty years from now, I think we're going to look back on a lot of this diagnostic stuff and say, "That was garbage." So as I see it, we have a choice. We can wait twenty years and several more editions of the *DSM* before we start to clean up this mess. Or we can take advantage of the technological resources that are beginning to become available and start phase three right now.

As you will soon see, I choose phase three.

6 Knowing Your Own Strengths

A FEW YEARS AGO Michelle Dawson, an autism researcher at the Rivière-des-Prairies hospital at the University of Montreal, asked herself an important question. Her research on the autistic brain, like the other autism research at the clinic, like autism research everywhere, focused on cognitive impairment — on what was wrong. And she realized that when an autistic person exhibited characteristics that we would call strengths if they belonged to a normal person, we still saw those strengths as merely the fortunate byproducts of bad wiring. *But what if they're not?* she asked herself. What if they're not the *byproducts* of anything? What if, instead, they're simply the *products* of wiring — wiring that's neither good nor bad?

She and her colleagues began digging in the literature. Sure enough, they found that studies routinely emphasized only the negative aspects of autism, even when some of the results were positive. According to Laurent Mottron, Dawson's frequent collaborator and the director of the autism program at Rivière-des-Prairies hospital, "Researchers performing fMRI scans systematically report changes in the activation of some brain regions as deficits in the autistic group — rather than evidence simply of their alternative, yet sometimes successful, brain organization." When researchers look at cortical volumes, for example, they automatically throw variations into the deficit bin, regardless of whether the cortex is thinner or thicker than expected. And even when a study does recognize a strength in autistic subjects, the authors often regard it as the brain's way of compensating for a deficit — but a 2009 report in the *Philosophical Transactions of the Royal Society* that reviewed papers based on this assumption concluded "that this inverse hypothesis rarely holds true."

Dawson and her colleagues began conducting their own experiments to determine the intelligence level of people with autism. In 2007 they designed a study that used two common tests of intelligence, the Wechsler Intelligence Scale for Children and the Raven's Progressive Matrices. The Wechsler test consists of twelve subtests, some verbal and some nonverbal (arranging blocks into designs, for instance). The Raven's is totally nonverbal. It consists of sixty questions that show a series of geometric designs and then a choice of six or eight alternative designs, only one of which completes the series. These tests were administered by independent neuropsychologists who were unaware of the purpose of the study, and the test subjects included fifty-one children and adults with autism and forty-three children and adult controls.

The results were striking. Dawson found that the measure of intelligence in the autistic population depended on the type of test. On the Wechsler, one-third of the test subjects with autism qualified as "low functioning." On the Raven's, however, only 5 percent did so — and one-third qualified as having "high intelligence." On the Wechsler, the autistic subjects on the whole scored far below the population average, while on the Raven's they scored in the normal range. I myself have scored really well on the Raven's Coloured Progressive Matrices.

Why such a wide disparity in responses to the two tests? Perhaps because answering many of the Wechsler questions correctly depends on the social ability to acquire skills and information from others, whereas the Raven's is purely visual.

"We conclude," the Montreal group wrote in their groundbreaking study published in *Psychological Science* in 2007, "that intelligence has been underestimated in autistics."

"Scientists working in autism always reported abilities as anecdotes, but they were rarely the focus of research," one of the authors of the paper, Isabelle Soulières, later said. "Now they're beginning to develop interest in those strengths to help us understand autism."

This new attitude toward autism is consistent with the phase-three thinking that I described in the previous chapter. Just as we can now begin to look at autistic-like behavior on a trait-by-trait basis, we can also reconceive autistic-like traits on a brain-by-brain basis.

Don't get me wrong. I'm not saying that autism is a great thing and all people with autism should just sit down and celebrate our strengths. Instead, I'm suggesting that if we can recognize, realistically and on a case-by-case basis, what an individual's strengths are, we can better determine the future of the individual. *I need you to fix me,* Carly Fleischmann, the nonverbal we met in chapter 4, once typed. *Fix my brain.* By contrast, when a journalist asked Tito Rajarshi Mukhopadhyay, the other nonverbal we met in chapter 4, "Would you like to be normal?" Tito answered, "Why should I be Dick and not Tito?" For Tito, the "acting self" might have been weird, but it was no less a part of him than his "thinking self."

I also want to be clear that when I say *strengths,* I'm not talking about savant skills like those of Stephen Wiltshire, who needs only one helicopter tour of a portion of a city, like London or Rome, in order to draw the entire landscape down to the last window ledge, or Leslie Lemke, who needs to hear a piece of music only once—any style, including complex classical compositions—in order to re-create it on the piano. Only about 10 percent of autistics belong in the savant category (though most savants are autistic).

So what strengths *can* we look for? While autism researchers traditionally haven't seen this trait as a strength, they've nonetheless noted over the years that people with autism often pay greater attention to details than neurotypicals. Let's start there and see where it takes us.

Bottom-Up Thinking

People with autism are really good at seeing details. "When a person with autism walks into a room," one researcher said, "the first thing they see is a stain on the coffee table and 17 floor boards." That seems an exaggeration and an overgeneralization to me, but the idea is on the right track.

Traditionally researchers have characterized this trait as "weak central coherence" — a deficit. Weak central coherence is at the heart of the impairments in social communication and social interactions that have long been part of the official diagnosis of autism. More informally, you can say that autistic people have trouble putting together the big picture, or that they can't see the forest for the trees.

Think about Tito and his encounter with the door. He saw the door as a series of properties — its physical features (hinges), its shape (rectangular), its function (allowing him to enter the room). Only when he'd collected enough details did he know what he was seeing. When I met him at a medical library, I asked him to describe the room. Rather than discussing the objects in the room or the size of the room, he talked about fragments of color.

My experience is nowhere near as extreme, but the tendency to see details before I see the big picture has always been a central feature in how I relate to the world. When I was a child, my favorite repetitive behavior was dribbling sand through my hands over and over. I was fascinated with the shapes; each grain looked like a tiny rock. I felt like a scientist working with a microscope.

A landmark study in 1978, "Recognition of Faces: An Approach to the Study of Autism," brought the social implications of this trait to the forefront of research. Subjects were shown only the lower parts of a series of faces of people they knew and asked to identify the people. The autistic population performed better than the controls. The same was true when both groups were shown inverted images. The people with autism were better at figuring out what the image was when it was upside down. The researcher who performed the study, Tim Langdell, posited that people with autism were better at seeing "pure pattern" rather than "social pattern."

This interpretation would be consistent with results from biological motion tests. You know motion-capture technology in filmmaking, where an actor wears a bunch of white dots that map his movements in a computer? That's biological motion. On a computer screen, biological motion is nothing more than dots moving, but the dots are arrayed in such a way that they suggest an action a living person or animal would perform, like running. Studies have repeatedly shown that people with autism can identify biological motion, but they don't do so with the same ease as neurotypicals. Nor do they attach emotions and feelings to the motions. What's more, they use different parts of the brain than neurotypicals do. Neurotypicals show a lot of activation in both hemispheres, while autistics show less activation overall. The way the autistic brain engages with biological motion is reminiscent of Tito's description of focusing on a door at the expense of seeing the room, or a description by Donna Williams I once read, of her being entranced by individual motes of dust.

The interpretation of this tendency as a deficit in social pattern recognition was adopted by R. Peter Hobson in an influential series of studies he spearheaded in the 1980s at the Institute of Psychiatry in London. Did children with autism prefer to sort photographs according to facial expressions exhibited (happy or sad) or the type of hat worn (floppy or woolen)? The hats won. Did children with autism have trouble putting the pieces of a face together into an interpretation of facial emotions? Yes.[1]

---

1. I myself didn't know that people have subtle eye signals until I was fifty. I have so much

These are important findings. But there can be a flip side to a deficit in social pattern recognition: a strength in pure pattern recognition — being really good at seeing the trees. Studies have repeatedly shown that people with autism perform better than neurotypicals on embedded-figure tests — a variation on the old something's-hidden-in-the-picture game. Several years ago I took a test where I had to look at large letters that were composed of smaller, different letters — for instance, a giant letter *H* that was built out of tiny *F*s. I then had to identify either the big letter or the little letter. I was faster at identifying the little letters, a result that's far more common among autistics than neurotypicals. Research has also shown that when performing language tasks, the autistic subject relies on the visual and spatial areas of the brain more heavily than the neurotypical subject does, perhaps to compensate for a lack of the kind of semantic knowledge that comes with social interaction. An fMRI study in 2008 showed that when the neurotypical brain conducted a visual search, most of the activity was confined to one region of the brain (the occipitotemporal, which is associated with visual processing), while what lit up in the autistic brain was just about everything. Perhaps this is why I can immediately spot the paper cup or hanging chain that's going to spook the cattle, while the neurotypicals all around me don't even notice it. Researchers have a lovely term for that tendency to see the trees before recognizing the forest: *local bias.*

Consider Michelle Dawson, the researcher who thought to look for references to autistic strengths that are buried in the literature. She's autistic. I can't say she made her conceptual leap *because* she's autistic, but I think she was more likely to make it because she herself possessed a fine attention to details. "Dawson's keen viewpoint keeps the lab focused on the most important aspect of science: data," Mot-

---

trouble remembering faces that in a business meeting, for instance, I'll force myself to recognize physical details: *Okay, she's wearing big glasses with black rims. He's the one with the goatee.*

tron wrote in a 2011 article in *Nature.* "She has a bottom-up heuristic, in which ideas come from the available facts, *and from them only.*"

Dawson had always approached her research with the same received wisdom, making the same unthinking assumption, as her mentors and peers — that studying autism means studying deficits. But that assumption was the result of what Mottron identified in himself as a "top-down approach: I grasp and manipulate general ideas from fewer sources." Only when he's come up with a hypothesis does he "go back to facts." Dawson found it easier to free herself of the preconceptions inherent in top-down thinking because she was able to see the details dispassionately and in isolation. When other researchers look at her data about autistic strengths and say, "It's so good to see something positive!" she answers that she doesn't see it as positive or negative: "I see it as accurate."

I completely identify with this attitude. For my undergraduate honors thesis, I wanted to explore the subject of sensory interaction. How does a stimulus to one sense, such as hearing, affect the sensitivity of other senses? I gathered more than one hundred journal papers. Because my thinking is totally nonsequential, I had to develop a way to make sense of the research.

First, I numbered each journal article. Next, I typed the major findings of each study on separate slips of paper. Some studies yielded only one or two strips of paper. Review articles prompted more than a dozen. Then I put all the strips in a box. I'd hung a huge bulletin board in my dorm room — maybe four feet by six feet. I drew the first strip out of the box and I pinned it just anywhere on the board. Then I pulled out the next strip. Let's say the first strip was about the sense of vision, and the second was about the sense of hearing. So the second strip went on a different part of the board, because now I had the beginnings of two categories. I made labels for those two categories and pinned them to the top of the board so that they headed two columns. I continued to take strips of paper out of the box, one at a time, like drawing lots. I'd pick one, put it with the other strips in a category, create a new category, or throw out all the old categories and rearrange all the strips of paper. In the end, after I had finished sorting all the strips of paper into different categories of information, I began to see how the categories of information fit together to form larger concepts.

I later applied this principle in my professional life. When I began to develop my livestock-handling designs, I first went to every feed yard in Arizona — maybe twenty — and then to a bunch of feed yards in Texas. I worked cattle in about thirty feed yards altogether, but what I was actually doing was observing. I would notice that one feed yard had a really nice curved lead-up chute and another had a nice loading ramp but terrible sorting pens. When I sat down to draw a design, I threw out all the bad bits and kept all the good bits.

This process can be extremely time-consuming. When I was in college, it sometimes took me months of reading journal articles and posting scraps of paper on the bulletin board to arrive at the basic principle. Now I have a lot more experience sifting through scientific research. I no longer need an actual bulletin board on the wall, because I've got one in my mind. That's why I trust my conclusions. I feel that my local bias frees me from the *global bias* that gets in the way of top-down thinkers.

Mottron identified the same pattern in Dawson's research. "She does need a very large amount of data to draw conclusions," he wrote in *Nature*. But, he added, "her models never over-reach, and are almost infallibly accurate."

This feeling of certainty is probably what has fed the reputation among mathematicians and scientists who have Asperger's or are high-functioning autistics as being rigid and unswerving. Once they arrive at a proof, their attitude toward it becomes inflexible, because they have experienced the piece-by-piece, painstaking logic that went into creating it. Mathematicians and scientists even speak of the beauty of an equation or proof.

For a top-down thinker, however, that certainty is *not* necessarily earned — not without a lot of supporting evidence. I had one client who insisted that he could build a meatpacking plant in three months. Well, no. That's absolutely not going to work. But he couldn't be persuaded otherwise. He *knew* he was right, and all the deadlines the contractor missed because they were impossible to meet, all the unforeseen delays that normally get padded into the schedule in advance, meant nothing. In the end, his was a twenty-million-dollar screwup.

For a bottom-up thinker like me, however, getting a detail wrong when I'm trying to solve a problem doesn't have implications for the whole solution, because I haven't reached the whole solution yet. If someone shows me a part of a project where I did something wrong, I say, "Change it."

Associative Thinking

Not long ago I was walking through the United Airlines terminal in Chicago, which has a glass roof. I looked up, and in my mind I saw pictures of the greenhouse at my university, the Crystal Palace from the 1851 World's Fair in London, a botanical garden, and the Biosphere in Arizona. These structures weren't the same shape as the airline terminal, but they were all in my glass-roof file.

Then when I saw the Biosphere in my mind, I noticed the turrets in the structure. They reminded me of the turrets on the Hoover Dam. So I started seeing pictures of turrets: on a castle in Germany, on the Disney Fantasyland castle, on a military tank.

At that point, I could have gone either way. I could have continued to root around in my glass-roof file. Or I could have stayed in the turret file. To an outsider, my thoughts might appear random, but to me, I'm simply selecting which file folder I want to explore.

I've often said that my brain works like a search engine. If you ask me to think about a certain topic, my brain will generate a lot of hits. It can also easily make connections that will get off the original topic

O'Hare's United terminal (left) and the Crystal Palace from the 1851 Great Exhibition in London.

© Ian Hamilton / Alamy (left); © Lordprice Collection / Alamy (right)

pretty fast and go pretty far. The similarity between my brain and a search engine, though, shouldn't come as too much of a surprise. Who do you think designed the original search engines? Very likely it was people whose brains work like mine — people with brains that have trouble with linear thought, brains that ramble, brains that have weak short-term memory.

Remember the HDFT scan of my brain at the University of Pittsburgh in 2012? It revealed that my corpus callosum — the neural highway that stretches the length of the brain between the left and right hemispheres — has an unusual number of horizontal fibers branching off to either side. My fibers bunch up back in the parietal area, which is associated with memory. I think all those extra circuits in the parietal area of my brain might well be what allows me to make a lot more associations than people with normal brains. "Oh," I said when Walter Schneider showed me images from the scan, "you found my search engine!"

Still, in order for a search engine to come up with hits, the database needs to be full of information to hit upon. In human terms, it needs memories.

Part of what made Michelle Dawson such a formidable researcher and collaborator, Mottron said, was that she possessed an exceptional memory: "Most nonautistic people can't remember what they read ten days ago. For some autistics, that's an effortless task. Autistic people are also less likely to misremember data."

Is that true? Is long-term memory generally better in people with autism?

I know that my *short-term* memory is horrible, which isn't unusual among high-functioning autistics. We're not good at multitasking. We have poor memories for faces and names. And sequencing? Forget it. A 1981 study showed that high-functioning children with autism remembered significantly less about recent events than normal age-matched and mentally handicapped age- and ability-matched control subjects. In a 2006 study of thirty-eight high-functioning autistic children and thirty-eight controls, the most reliable and accurate test to distinguish between the two groups was the Finger Windows subtest—a measure of spatial working memory in which the experimenter touches a series of pegs on a board and the subject has to duplicate the pattern sequence. The controls easily outperformed the high-functioning autistics. When I took this test, I trashed it; it placed too much of a workload on my working memory.

But what about long-term memory in people with autism? To my surprise, the scientific literature in this area is exceedingly thin. I spent two hours searching the Internet for peer-reviewed papers on the topic; the most recent was from 2002, and it was basically asking if long-term memory was *impaired* in autistics.

Still, whether long-term memory in autistic people tends to be better or worse than it is among neurotypicals is almost beside the point. The fact is, you need memories. You need data.

When I was in college looking at my bulletin board, I didn't have a lot of experience in research, and because I was still relatively young, I didn't have a lot of experience in life. As I've turned forty, then fifty, then sixty, my ability to make associations — to see connections between details — has become more and more acute, and my need to use a bulletin board has disappeared, because I have more and more details in my database. Think of it this way: If you can't see the trees, you'll never see the forest.

Creative Thinking

The forest that the autistic brain winds up seeing, however, might not look the same as the forest that the neurotypical brain sees.

I recently read a definition of *creativity* in the journal *Science* that really made an impression on me: "a sudden, unexpected recognition of concepts or facts in a new relation not previously seen." That's what happened when Michelle Dawson challenged the whole history of autism research based on identifying deficits. She had the same concepts and facts as everyone else, but she saw them in a "new relation not previously seen."

I can think of plenty of examples of this sort of creativity from my own life. I remember when I was a student at Franklin Pierce College, and I took a course on genetics. The professor, Mr. Burns, taught us the usual model of genetics developed by Gregor Mendel in the nineteenth century — that each parent contributes half the genes in an offspring and that the way species gradually change is through a long series of random genetic mutations. That didn't make sense to me. Sure, it was part of the explanation. But it couldn't be the whole explanation. How do random mutations explain that when you take a Border collie and a springer spaniel and breed them, you get puppies that look like they're a mixture of the two breeds, but they're not *exactly* half and half? Some puppies look more like spaniels, and others look more like collies. I actually went up to Mr. Burns and asked him: "How does Mendel explain that?"

He was surprised, to say the least. But today we know that random mutations are not enough to produce the diversity within species. Evolution also needs copy number variations. What Mendel's genetics tells you is that you have genes. But the concept of copy number variations tells you that you have either a lot of copies or just a handful.

A few years ago I went to a reunion at Franklin Pierce and I saw Mr. Burns, who was by then retired. "You asked some questions that were really deep," he told me. They didn't seem deep to me. They seemed like common sense. But now I understand I wouldn't have been able to make the association between Mendel's genetics and crossbred dogs if I didn't already have enough crossbred dogs in my database. In fact, when I confronted Mr. Burns, I had in mind a particular Border collie and springer spaniel that I had known back when I was in high school. They were the parents of a litter of puppies. I could still see the mom and dad in my mind, and I could see the puppies, and I could see what the dogs looked like when they grew up.

I like to look at the usual materials for any project and imagine a potential application or construction that wouldn't occur to most other people. I wouldn't say that all autistic people are creative, or that creativity is a happy byproduct of autism. Whole-genome studies have indicated some de novo copy number variation overlap between autism and schizophrenia, and highly creative people have demonstrated elevated risk for schizophrenia and other psychopathologies. This area of research, however, is still preliminary. But I will say that I think being autistic makes a certain *kind* of creativity more likely to arise. To illustrate what I mean, I'm going to show you a test I recently took.

The challenge in this test, which originally appeared in a brain study and was reproduced in *New Scientist,* was to use a circle to create as many drawings as possible in five minutes. That's all the illustration showed: a simple circle. The two examples the article gave were a smiley face, which was "among the most unoriginal," and a man reclining in an airplane seat (so that the circle was the porthole window, looking into the plane from the outside).

The drawings I produced were:

1. The rifle target iris from the opening credits of James Bond movies
2. Camera iris
3. Bike wheel
4. Periscope image of a boat
5. Round bison facility (which I had actually designed)
6. Merry-go-round (seen from above)
7. Rotating milk parlor

At this point, I began to wonder about the ground rules. Could I go outside the circle? I drew a:

8. Ferris wheel, with seats swinging from the circle

I wasn't sure if that drawing was legal, but what the heck. I was on a roll. So I drew a:

9. Hamster wheel — with a base, so that it wouldn't fall over

Then I wondered if I could use the circle as the center of a larger drawing, in which case I could draw all sorts of flowers.

This test is a variation on an old classroom exercise I often use; let's call it Thinking Outside the Brick. I ask, "How many uses can you think of for a brick?" Right away I'll get the obvious answers. You can use it to build a wall. You can throw it through a window. It usually takes the students a while (with the help of a hint or two from me) before they realize that they can change the form of the brick. You can grind it up and use it as a pigment in paint. You can chop it up into cubes, paint dots on the cubes, and play dice.

The trick to coming up with novel uses for a brick is to not be attached to its identity as a brick. The trick is to reconceive it as a non-brick.

I think that bottom-up, details-first thinkers like myself are more likely to have creative breakthroughs just because we don't know where we're going. We accumulate details without knowing what they mean and without necessarily attaching emotional significance to them. We seek connections among them without knowing where they're taking us. We hope those associations will lead us to the big picture — the forest — but we don't know where we will be until we arrive there. We *expect* surprises.

Earlier in this chapter, I mentioned that autistic people generally tend to see details better than neurotypicals, and then I said, "Let's start there and see where it takes us." It's taken us here: to a creative leap about creative leaps — specifically, that the autistic brain might be more likely, on average, to make a creative leap. An attention to details, a hefty memory, and an ability to make associations can all work together to make the unlikely creative leap ever more likely.

In his book *Be Different: Adventures of a Free-Range Aspergian,* John Elder Robison described this progression of creativity — one that led to his career creating sound effects and musical instruments and designing laser shows and video games. He wrote that he first became interested in music as an adolescent, because he was fascinated with the patterns that music waves made on an oscilloscope, a device that displays electric signs and lines and shapes on a small screen. "Each signal had its own unique shape," he wrote. These signals were the *bottom-up details.*

He spent eight to ten hours a day "absorbing music and unraveling how the waves looked, and how electrical signals worked," he wrote. "I watched and listened and watched some more until my eyes and ears became interchangeable." In other words, he was storing up *memories.*

"By then, I could look at a pattern on the scope and know what it sounded like, and I could look at a sound and know what it looked like." Based on those memories of details, he had taught himself how to make the necessary *associations.*

Then he was ready for the creative leap:

> If I set the scope to sweep slowly, the rhythm of the music dominated the screen. Loud passages would appear as broad streaks, while quiet passages thinned down to a single tiny squiggle. A slightly higher sweep speed showed me the big, heavy, slow waves of the bass line and the kick drum as wide squiggles. Most of the energy was contained in those low notes. Up higher, with a faster scope setting, I found the vocals. At the top of it all lay the jagged fast waves from the cymbals.
>
> Every instrument had a distinct pattern, even when they were all playing the same melody. With practice, I learned how to distinguish a passage played on an organ from the same music played on a guitar. But I didn't stop there. As I listened to the instruments, I realized each one had its own voice. *"You're nuts," my friends said, but I was right. The musicians all had their own ways of playing, but their instruments were unique, too.*

The emphasis is mine. The neurotypical response to his insight was to dismiss it. But Robison could hear what other people missed.

Actually, he could *see* it: "I saw the whole thing as a great mental puzzle—adding the waves from different instruments in my head, and figuring out what the result would look like." He was, he learned, working in a kind of waveform mathematics, even though he didn't think of his work as math.

Seeing waves, adding them in his head—that sounded like visual thinking, as in "thinking in pictures." That's my kind of thinking. But

I definitely didn't see the kinds of things Robison described. I saw concrete examples from my past, not abstractions. He and I both used our autistic brains to be creative, and the creativity was visual, yet his kind of creativity wasn't my kind of creativity.

In figuring out how to make the most of the autistic brain's strengths, I apparently still had at least one more creative leap to make.

7 Rethinking in Pictures

This is mostly good [sic], *well rounded book. However, Dr. Grandin does make some glaring over generalizations, and often seems to assume all autistic people are like herself. While she admits this is not true in place* [sic], *she will go to the very next paragraph and say something like, "because all autistic people are visual . . . ," when, in fact, some autistic people have severe visual processing difficulties and are not visual at all. While I can relate to most of what she says as an autistic person, I know many who cannot.*

LIKE MANY AUTHORS, I read reviews of my books on Amazon.com. This review, from 1998, was one of the first to appear for my book *Thinking in Pictures,* and I admit it really stung. I didn't exactly treat it like hate mail. I didn't think somebody was trying to hurt me. But I didn't take it lightly either. Some autistic people "are not visual at all"? Could it be true?

I wrote *Thinking in Pictures* because I had come to understand that the way I saw the world wasn't the way other people saw the world. Even after I learned that I'm autistic, I didn't think about whether autism affected the way I saw the world. When I started designing livestock facilities in the 1970s, I couldn't understand why other designers didn't see obvious mistakes—mistakes that I could see at a glance. I thought those people were stupid. Of course, I understand now that we were just looking at the world through very different sets of eyes — or, I should say, through very different kinds of brains. So I had been mistaken. Not everyone thinks in pictures? Okay. But people with autism do.

I'd had good reason to think that all autistic people were visual thinkers and only visual thinkers. As far back as 1982, when I was writing a paper that later appeared in the *Journal of Orthomolecular Psychiatry,* I came across several pieces of research that supported these assumptions. One study reported that autistic children scored normally on the Wechsler block design and object assembly tests. Another study reported that autistic children seemed to "perform badly on tests which require verbal or sequencing skills, even if the tests do not involve the use of speech." On the basis of this research and my own experience of seeing the world, I felt comfortable with my conclusion: "Studies of autistic children by many different researchers indicate the visual spatial nature of the autistic mind."

Well, I was right. That's what those studies did indicate. But what about that Amazon reviewer — and other Amazon reviewers who came along to echo the complaint?

Ever since the first review appeared, I've given a great deal of thought to the topic of different ways of thinking. We can reconceive the autistic brain as a repository for certain strengths — the ability to pick out details, maintain a large database of memories, make associations. But of course, autistic brains don't all see the world the same way — despite what I once thought. Autistic brains might tend to have these strengths in common, but how each individual uses them varies. What kinds of details? What kinds of memories? What kinds of associations? The answers to these questions depend on what type of thinker you are, because a brain that focuses on words is not going to reach the same conclusions as a brain that focuses on pictures.

In fact, my pursuit of this topic has led me to propose a new category of thinker in addition to the traditional visual and verbal. At this point, this third category is only a hypothesis. But it has transformed my thinking about autistic people's strengths. And I've even found scientific support for this hypothesis.

For years I had been giving lectures, and I had made an assumption without even knowing it: I think in pictures; I'm autistic; therefore, all autistic people think in pictures. Made sense to me. If you say the word *train* to me, I automatically see a subway train in New York; a train that goes right through the campus of the university where I teach; a coal train in Fort Morgan, near my home; a train I rode in England that was standing room only, full of soccer hooligans who took up all the seats and wouldn't let anybody else sit for the whole, miserable four-hour ride; a train in Denmark where kids teased me until the newsstand lady made them go away.

But now I wanted to find out whether the autistic people in the audience actually did think the same way I did. So I started asking the audience members who introduced themselves to me after my lectures, "What was"—or "is," if I was talking to a child—"your favorite subject in school?" Often the answer wasn't art class, as you would expect from a visual thinker. Instead, a lot of the time it was history.

History? I thought. *History is full of facts, and facts are full of words, not pictures.*

So, okay. People with autism can think in visual terms or verbal terms, just like neurotypicals. That Amazon reviewer was right.

But then one day in early 2001 I got an advance copy of a book in the mail, *Exiting Nirvana: A Daughter's Life with Autism*, by Clara Claiborne Park. The publisher wanted to know if I would write a blurb for it—a quote recommending the book that would appear on the back cover. I already knew about Clara and her daughter Jessica, or Jessy. Jessy was born about ten years after me, when the medical consensus about autism had shifted toward the psychoanalytic search for psychic wounds. Since Jessy is younger than me, Clara Park had to battle the medical establishment constantly in an effort to get people to understand that the source of her daughter's behavior wasn't in her mind. It was in her brain.

I had written about Jessy a little bit in *Thinking in Pictures*; I referred to a 1974 paper that examined the elaborate system of symbols and numbers that Jessy had invented in order to navigate her life. Things she considered very good, like rock music, she labeled with four doors and no clouds. Things she considered pretty good, like classical music, rated two doors and two clouds. And the spoken word deserved zero doors and four clouds — the worst rating.

So when I received that advance copy of *Exiting Nirvana,* I was eager to read it. What I found, though, shocked me.

I knew Jessy was an artist, but nothing prepared me for what I saw in this book. Her art was unlike anything I'd ever seen. It was full of psychedelic colors — vibrant, almost neon shades of orange and pink and turquoise and chartreuse and tangerine and plum. But she applied these to objects that would never have those colors. The cables of a bridge. The windows of an office building. The siding of a house.

What category did this kind of mind belong to? Visual or verbal? Visual, obviously. But that couldn't be the whole story, because I'm a visual thinker, and I sure didn't think like *that.*

She painted the objects in her artwork in photorealistic detail from memory, so she clearly could think in pictures, just as I do. But her artwork wasn't like my drawings; the pictures she saw in her mind weren't my kinds of pictures. When Jessy drew a building, the emphasis was on colors and patterns. When I drew a structure, the emphasis was on the details of the different surfaces — round pipes, concrete grooving, metal gratings. Jessy might have had files full of images in her mind, just like me, but she could manipulate those images in ways I couldn't begin to imagine.

So what kind of mind *was* hers? How was her brain wired? Did my system of dividing the world of autism into picture thinkers and word-fact thinkers deserve a rating of zero doors and four clouds?

I stopped looking at the pictures and started reading. I focused

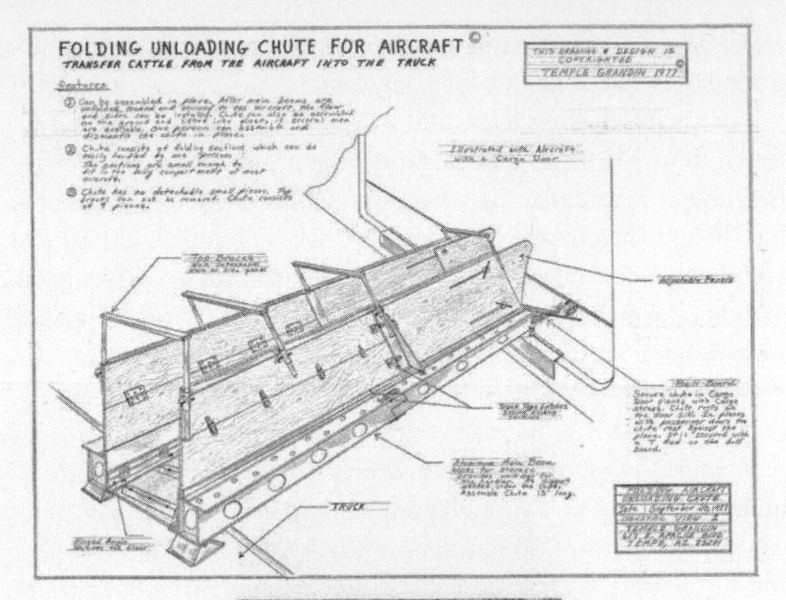

In black and white, you can see that my idea of 3-D and Jessy's idea of 3-D look very similar in their attention to mechanics and detail. But go to the Internet to see what black and white can't capture in Jessy's work: a vibrant mosaic of colors. *© Temple Grandin (top); © Jessy Park (bottom), photo courtesy of Pure Vision Arts*

especially hard on anything that might give me a clue about Jessy's thinking. On page 71, I read that Jessy liked to search out regularities in words. "She thought about them, talked about them, wrote them down. Elf, elves; self, selves; shelf, shelves; half, halves" — et cetera. In the margin next to this paragraph I wrote *word patterns.*

On the following page, Jessy's mother, Clara, described a book that Jessy made shortly after her fourteenth birthday. It was, she wrote, "a celebration of the transformations of words. The book was a thing of beauty, a theme and variations, four words in three colors: SING, SANG, SUNG, and SONG."

At the bottom of the page, I wrote *word patterns.*

"Clocks became fascinating," Clara wrote about Jessy in the following chapter,

> when she learned that the French numbered time not in twelve hours but in twenty-four. She drew a ten-hour clock, a twelve-hour clock, a fourteen-hour clock, sixteen-, eighteen-, twenty-four-, and thirty-six-hour clocks. She converted hours to minutes, minutes to seconds; surviving sheets record that 3600 seconds = 60 minutes = 1 hour. Carefully she drew in each second. Time was now something to play with. Fractional conversions became so rapid as to seem intuitive: 49 hours = 2 1/24 days. Soon she was mapping space as well as time: 7½ inches = ⅝ foot.

Finds all the patterns, I scribbled in the margin.

Wait a second.

Patterns.

Three times I had used the word in the span of just a few pages.

I thought about the Raven's Progressive Matrices test. The subject is shown a pattern or matrix from which a piece is missing and then has to choose the piece that completes the puzzle. I knew from *Exiting Nirvana* that at the age of twenty-three, Jessy had scored in the ninety-fifth percentile on that test. Then she took the *Advanced* Progressive Matrices. Again she scored in the ninety-fifth percentile.

I also thought about a work of origami—the Japanese form of art that comes from the words for "folding" and "paper"—that a boy presented to me after one of my talks. It was unlike any work of origami art I'd ever seen. I had made origami figures myself, but I used just one sheet of paper for each and followed simple instructions that produced the most common origami designs, such as a crane. But this boy's origami was full of colors, each color belonging to a separate sheet, and the design was the shape of a star. I was so impressed that when I returned home from that trip, I gave the origami star a place of honor on a windowsill where I could see it every day. Sometimes I would take it down from the windowsill and study it.

The star was about three inches by three inches by three inches. It had eight points. Each point had three colors, and no two points had the same combination of colors. I tried to count the colors, but because of my poor working memory, I had to write them down in order to be sure I had counted them all. Pink, purple, red, light green, dark green, blue, yellow, orange. Eight colors, meaning eight sheets of paper. All of the sheets of paper were interlocking, and the base of each triangular point intersected with the bases of the other triangular points.

After the boy had presented me with his gift, he hurried away, but I noticed that his parents were still standing nearby. I asked them about their son, and they said he was gifted in math. Which made sense. It certainly took a mathematical mind to engineer such a complicated structure. But didn't such a subtle and beautiful work of art have to be the product of a visual mind too? *Maybe,* I thought one day, putting the origami back on the windowsill, *people who are really good at math think in patterns.*

Once I realized that thinking in patterns might be a third category, alongside thinking in pictures and thinking in words, I started seeing examples everywhere.

After I gave a talk at one high-tech firm in Silicon Valley, I asked some of the folks there how they wrote code. They said they actually visualized the whole programming tree, and then they just typed in the code on each branch in their minds. And I thought, *Pattern thinkers.*

I recalled my autistic friend Sara R. S. Miller, a computer programmer, telling me that she could look at a coding pattern and spot an irregularity in the pattern. Then I called my friend Jennifer McIlwee Myers, another computer programmer who is autistic. I asked her if she saw programming branches. No, she said, she was not visual in that way; when she started studying computer science, she got a C in graphic design. If someone gave her a verbal description, she said, she couldn't "see" it. When she read the Harry Potter books, she couldn't make sense of the Quidditch competitions; she didn't understand what was happening until she saw the movies. But, she said, she did think in patterns. "Writing code is like crossword puzzles, or sudoku," she said.

Crossword puzzles involve words, of course, while sudoku involves numbers. But what they have in common is pattern thinking. In the 2006 documentary *Wordplay,* a movie about crossword puzzles, the people who created the best puzzles were mathematicians and musicians. And improving your sudoku-solving skills requires a greater and greater awareness of the patterns in the grid.

Then I read an article on origami in *Discover* magazine that just about blew my mind. I learned that for hundreds of years, the most complex origami patterns needed only about twenty steps, but in recent years, competitors in extreme origami had used software programs to design patterns requiring one hundred steps. And I read this amazing passage:

> The reigning champion of intricate origami is a 23-year-old Japanese savant named Satoshi Kamiya. Unaided by software, he recently produced what is considered the pinnacle of the field, an eight-inch-tall Eastern

> dragon with eyes, teeth, a curly tongue, sinuous whiskers, a barbed tail, and a thousand overlapping scales. The folding alone took 40 hours, spread out over several months.

How did he perform such a feat? "I see it finished," he said. "And then I unfold it in my mind. One piece at a time." *Patterns.*

In 2004, Daniel Tammet came to my, and a lot of other people's, attention when he set a European record for reciting the highest number of digits of pi ever: 22,514. And he did so in five hours. That's an average of 75 digits a minute — more than one per second. Demonstrations of other abilities followed: He became fluent in Icelandic in only a week; he could tell you what day of the week a distant date would fall on. In interviews, he said that he had been diagnosed with Asperger's syndrome. When he published his book *Born on a Blue Day,* I naturally couldn't wait to read it.

He explained the title on page 1: He was born on January 31, 1979, a Wednesday — and Wednesdays, in his mind, were always blue. As I read on, I learned that he thought of numbers as unique, each having its own personality. He said that he had an emotional response to every number up to 10,000. He described seeing numbers as shapes, colors, textures, and motions. He explained that he could instantly multiply two large numbers — 53 × 131, for example — not by performing the math but by "seeing" how the shapes of the numbers merged into a new shape, which he recognized as the number 6,943.

Patterns.

I wanted to know more about his thinking, so I found an interview in which he discussed how he learned languages. When teaching himself German, for instance, he noticed that "small, round things often start with 'Kn'" — *Knoblauch* (garlic), *Knopf* (button) and *Knospe* (bud). Long, thin things often start with "Str," like *Strand* (beach), *Strasse* (street), and *Strahlen* (rays). He was, he said, looking for *patterns.*

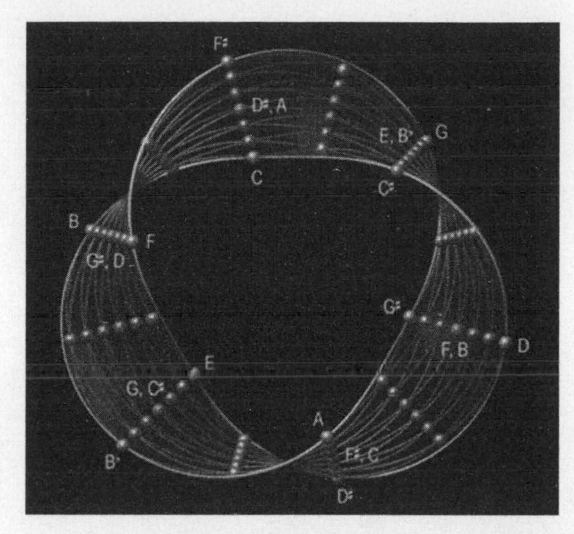

Music as Möbius strip.
© Rachel Hall

Now, I'm certainly not the first person to notice that patterns are part of how humans think. Mathematicians, for instance, have studied the patterns in music for thousands of years. They have found that geometry can describe chords, rhythms, scales, octave shifts, and other musical features. In recent studies, researchers have discovered that if they map out the relationships between these features, the resulting diagrams assume Möbius strip–like shapes.

The composers, of course, don't think of their compositions in these terms. They're not thinking about math. They're thinking about music. But somehow, they are working their way toward a pattern that is mathematically sound, which is another way of saying that it's universal. The math doesn't even have to exist yet. When scholars study classical music, they find that a composer such as Chopin wrote music that incorporated forms of higher-dimensional geometry that hadn't yet been discovered. The same is true in visual arts. Vincent van Gogh's later paintings had all sorts of swirling, churning

In 1889, Vincent van Gogh arrived at a visual representation of a *Starry Night* that matched the mathematics of turbulent flow—a formula that wasn't discovered until the 1930s.

patterns in the sky—clouds and stars that he painted as if they were whirlpools of air and light. And, it turns out, that's what they were! In 2006, physicists compared van Gogh's patterns of turbulence with the mathematical formula for turbulence in liquids. The paintings date to the 1880s. The mathematical formula dates to the 1930s. Yet van Gogh's turbulence in the sky provided an almost identical match for turbulence in liquid. "We expected some resemblance with real turbulence," one of the researchers said, "but we were amazed to find such a good relationship."

Even the seemingly random splashes of paint that Jackson Pollock dripped onto his canvases show that he had an intuitive sense of patterns in nature. In the 1990s, an Australian physicist, Richard Taylor, found that the paintings followed the mathematics of fractal geometry—a series of identical patterns at different scales, like nesting Russian dolls. The paintings date from the 1940s and 1950s. Fractal geometry dates from the 1970s. That same physicist discovered that he could even tell the difference between a genuine Pollock and a forgery by examining the work for fractal patterns.

"Art sometimes precedes scientific analysis," one of the van Gogh researchers said. Chopin wrote the music he wrote, and van Gogh and Pollock painted the images they painted, because something just felt right. And it just felt right because, in a sense, it *was* right. On some deep, intuitive level, these geniuses understood the patterns in nature.

And the relationship between art and science can go the other way too; scientists can use art to understand math. The physicist Richard Feynman revolutionized his field in the 1940s when he devised a simple way to diagram quantum effects: A straight, solid line represented particles of matter or antimatter, which traveled through space and time. Wavy or dashed lines represented force-carrying particles. When an electron moving in a straight line emitted a photon in a wavy line, the straight line recoiled to the right. Equations that

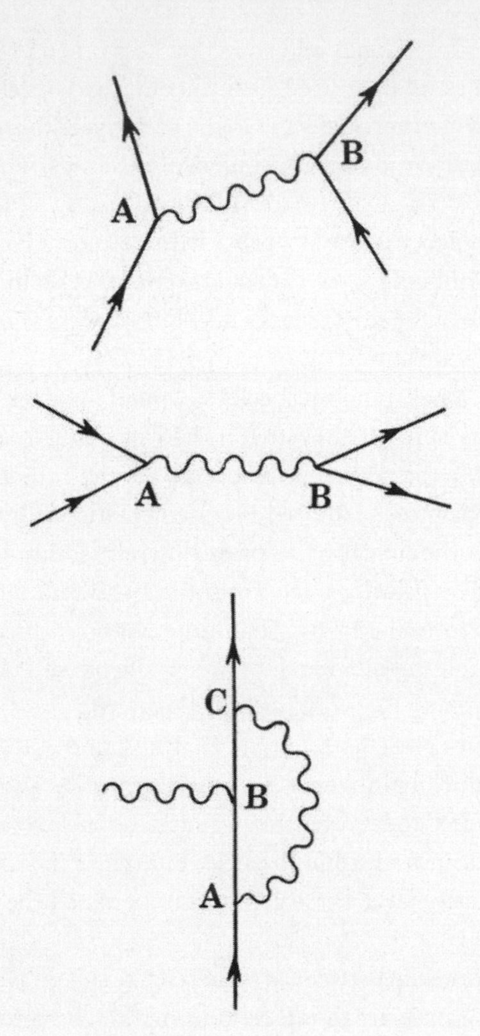

Richard Feynman taught physicists a new way to "see" quantum effects simply by deploying straight lines and wavy lines. From top to bottom: a muon at A kicking an electron at B out of an atom by exchanging a photon (wiggly line); an electron and positron annihilating at A and producing a photon that rematerializes at B as new forms of matter and antimatter; an electron emitting a photon at A, absorbing a second photon at B, and then reabsorbing the first photon at C.

© SPL / Photo Researchers, Inc.

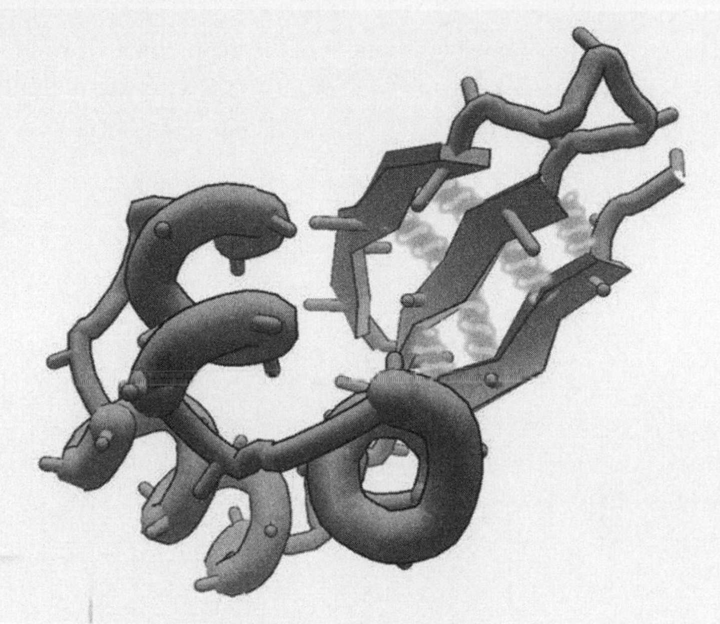

A Foldit solution to the crystal structure of M-PMV retroviral protease by molecular replacement—as discovered by nonscientists using their pattern thinking. © *University of Washington Center for Game Science*

took months to calculate could suddenly be understood, through diagrams, in a matter of hours.

In 2011, participants in an online video puzzle game called Foldit solved the mystery of the crystal structure of a particular monomeric retroviral protease. The configuration of the enzyme had long eluded scientists, and the solution was so significant that it actually merited publication in a scientific journal. What made the achievement especially remarkable, though, was that the players were not biochemists. But they sure were pattern thinkers.

Mathematicians distinguish subsets of thinkers: algebra thinkers and geometry thinkers. Algebra thinkers look at the world in terms of numbers and variables. Geometry thinkers look at the world in terms of shapes. Do you remember the Pythagorean theorem? It's this: The sum of the squares of the lengths of the legs of a right triangle is equal to the square of the length of the hypotenuse.[1] If you're an algebra thinker, you see $a^2 + b^2 = c^2$. But if you're a geometry thinker, you see:

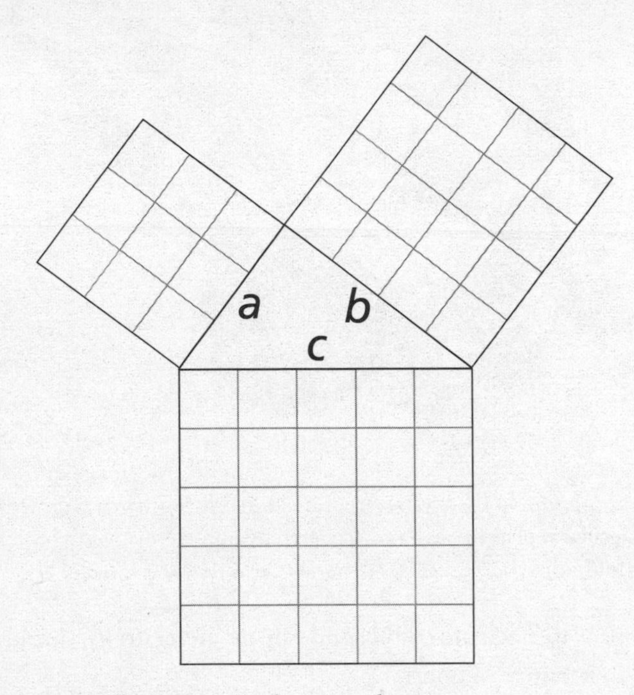

© *Houghton Mifflin Harcourt / Academy Artworks*

And then there's chess. There's always chess. For a century now, chess has been the petri dish of choice for cognitive scientists — researchers who think about thinking. Skill at chess can easily be measured, which is why rankings can be so precise, and it can be observed in an environment as controlled as any laboratory's — the tournament hall.

1. By the way, pay no attention to the Scarecrow in *The Wizard of Oz* after he receives his brain. What he apparently intends to recite is the Pythagorean theorem. What he actually recites is: "The sum of the square roots of any two sides of an isosceles triangle is equal to the square root of the remaining side" — which is gibberish. Poor Scarecrow.

What makes a chess master a chess master? Definitely not words. But not pictures, either, which is what you might think. When a chess master looks at the board, she doesn't see every game she's ever played and then find the move that matches the move from a game she played three or five or twenty years earlier. (That's probably what I would try to do.) A chess master doesn't "see" a board from a nineteenth-century chess match that she's studied closely.

So what *does* a chess master see, if not pictures? By now you can probably guess: patterns.

The stereotype of a chess grand master is someone who can think many moves ahead. And certainly, many chess players do strategize that way. Magnus Carlsen, a Norwegian prodigy who became a grand master in 2004 at the age of thirteen, calculates twenty moves ahead and routinely makes moves that other grand masters haven't even contemplated. Most grand masters can see many moves ahead, even while playing dozens of games simultaneously, walking from board to board in an exhibition space.

But a clue to how they're thinking comes from José Raúl Capablanca, a Cuban grand master. In 1909, he participated in an exhibition in which he played twenty-eight games at once, and he won all twenty-eight. His strategy, though, was the opposite of Magnus Carlsen's.

"I see only one move ahead," Capablanca reportedly said, "but it is always the correct one."

Cognitive scientists don't see a contradiction between these two approaches. Whether a chess player immediately sees a move in the context of twenty moves ahead or immediately sees a move in the context of one move ahead, the point is that he sees the move *immediately.*

The grand masters see it immediately not because they have better memories than regular players. They don't, according to studies that tested their memories. Nor do masters and grand masters see the next move immediately because their memories carry more possibilities from which they can choose. Their memories *do* carry more possibilities, because top-tier players have played longer than other players. But they retrieve from their memories not *more* possibilities but *better* possibilities. It's not just the quantity that grows over time. It's the quality.

But even having access to higher-quality moves doesn't explain why top players can see their next moves immediately. The reason is that they are better at recognizing and retaining *patterns*—or what cognitive scientists call *chunks.*

A chunk is a collection of familiar information. The letter *b* is a chunk, as is the letter *e,* as is the letter *d.* The ordering of those letters as *bed* is also a chunk, as is the phrase *going to bed.* The average person's short-term memory can retain only about four to six chunks. When superior chess players and novices were presented with pieces on nonsensical boards and then asked to re-create the positions of the pieces from memory, members of both groups were able to recall the locations of four to six pieces. When they were presented with pieces on actual chessboards, however, the superior chess players could recall the positions of the pieces across the board, while the novices were stuck at the four-to-six-pieces level. The real-life chessboards contained familiar patterns of pieces, and for a superior player, each pattern represented a chunk. To the expert eye at a glance, a board of twenty-five pieces might have four or six chunks—and the master or grand master knows upward of fifty thousand chunks, which is to say upward of fifty thousand patterns.

Michael Shermer, a psychologist, historian of science, and professional skeptic (he founded *Skeptic* magazine), called this property of the human mind *patternicity.* He defined *patternicity* as "the tendency to find meaningful patterns in both meaningful and meaningless data." Why would we need to find patterns even when they're not there? "We can't help it," he wrote in his book *The Believing Brain.* "Our brains evolved to connect the dots of our world into meaningful patterns that explain why things happen."

In fact, we might make bad decisions because our brains themselves feed us bad information. Our brains "want" to see patterns, and as a result, they might identify patterns that aren't there. In one experiment, for instance, researchers found that when subjects were shown randomly pointing lines on a computer screen and were asked which way, on average, the lines were pointing, they consistently tended to think the lines were pointing in either a more horizontal or a more vertical way than they actually were. The researchers hypothesized that our brains "want" to see horizontal or vertical, because that's what we need to see in nature. The horizon tells us where we're headed; the vertical tells us there's an upright person coming our way.

Even if the ability to identify patterns in nature isn't foolproof, it *is* exquisitely calibrated, and without it we wouldn't be here. It is as fundamental a part of our thinking as images and words. Patterns seem to be part of who we are.

Think of the golden ratio: Take a line and divide it into two unequal segments. If the ratio of the line's total length to the length of the longer segment is the same as the ratio of the length of the longer segment to the length of the shorter segment, then the two segments are said to be in the golden ratio. That number, rounded, is 1.618, and for thousands of years, mathematicians have pondered its "ubiquity

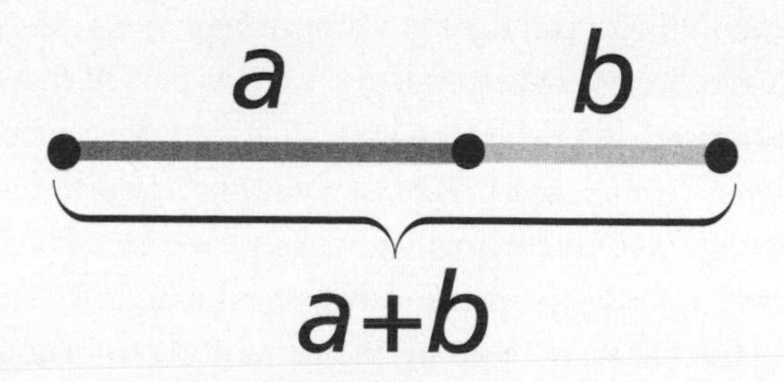

The golden ratio: The ratio of the whole length (a + b) to the longer of the two sections (a) is the same as the ratio of the longer of the two sections (a) to the shorter (b). © *Hougton Mifflin Harcourt / Margaret Anne Miles*

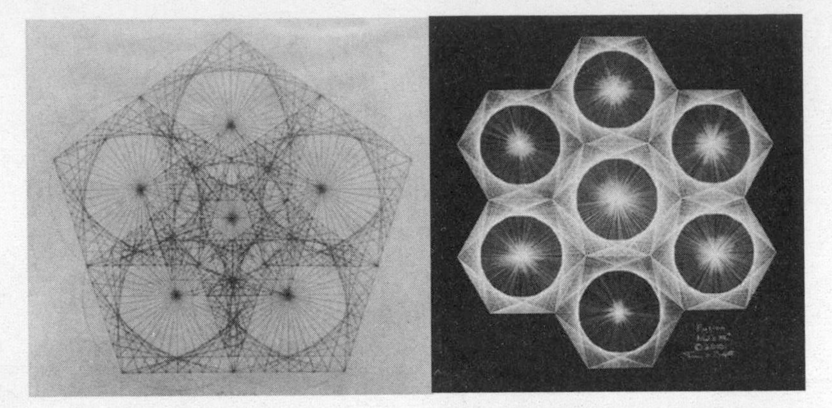

Jason Padgett's fractal art: *Quantum Star* (left); *Blue Fusion* (right)
© Jason D. Padgett

and appeal," as the astrophysicist Mario Livio wrote in his book *The Golden Ratio.* "Biologists, artists, musicians, historians, architects, psychologists" have studied it, he wrote. "In fact, it is probably fair to say that the Golden Ratio has inspired thinkers of all disciplines like no other number in the history of mathematics."

About a decade ago, a college dropout named Jason Padgett survived a vicious mugging outside a karaoke bar in Tacoma, Washington. He was struck in the back of the head, just above the primary visual cortex, and he suffered a concussion. Then a day or two later, he began seeing the world as a mathematical formula. "I see bits and pieces of the Pythagorean theorem everywhere," he said. "Every single little curve, every single spiral, every tree is part of that equation." He found himself compelled to draw what he was seeing, over and over and over, year after year. All the resulting artwork turned out to be fractals that were mathematically precise—even though he had had no math training and previously had exhibited no talent for art. It's as if the fractals were in his brain, just waiting to be freed.

And maybe they were. Way back in 1983, I clipped a *New Scientist* article that considered this possibility. (I guess the subject of pat-

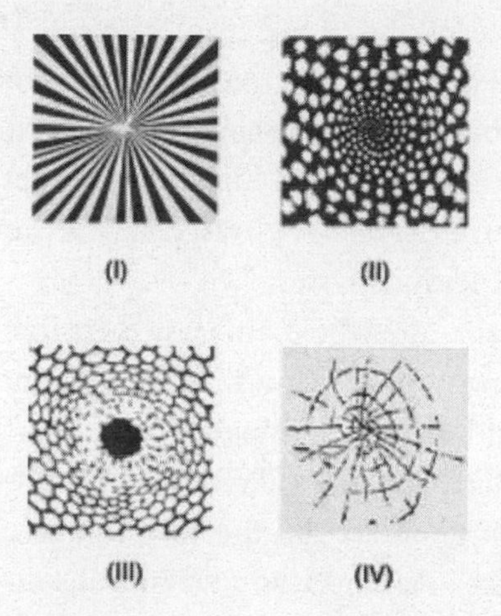

Heinrich Klüver's categorization of hallucinations: (I) tunnels and funnels, (II) spirals, (III) lattices, including honeycombs and triangles, and (IV) cobwebs. © *"Spontaneous Pattern Formation in Primary Visual Cortex," by Paul C. Bressloff and Jack D. Cowan*

terns was interesting to me even then, though I wouldn't realize it for nearly two decades.) The article concerned the research of Jack Cowan, a mathematician then at the California Institute of Technology, into visual hallucinations induced by drugs, migraines, flickering lights, near-death experiences, or any other catalyst.

In 1926, the German-born psychologist Heinrich Klüver noted that hallucinations fell into one or more of four basic categories: lattices, like checkerboards and triangles; tunnels or funnels; spirals; and cobwebs. "People have been reporting on this . . . ever since reported history, and even before," Cowan said in an interview. "You see it in cave paintings and rock art, and everybody seems to see the same kinds of imagery and it seems to be rather geometric."

Cowan hypothesized that because hallucinations moved independently of the eye, the source of the image was not on the retina but in the visual cortex itself. "What that told me," he said, "was that if you see geometric patterns, the architecture of your brain must be reflecting those patterns and therefore must itself be geometric."

Cowan and other researchers have continued to pursue that idea over the past three decades, and today they accept, as a 2010 review article in *Frontiers in Physiology* phrased it, "the prevalence of fractals at all levels of the nervous system."

You could say that the whole universe is fractal. Look at the weblike structure of neuronal cells in the brain, the network that transmits chemical and electrical signals. Then look at the large-scale structure of the universe, the galaxy clusters and superclusters that make up what astronomers call the cosmic web. If you squint, you can't tell them apart. Perhaps it should come as no surprise that cosmologists at the Johns Hopkins Institute for Data-Intensive Engineering and Science are trying to figure out the complexity in the evolution of the cosmic web by applying the principles of origami.

Still, I had to ask myself: Is there actually such a thing as a *pattern thinker*? Does pattern thinking deserve *a category of its own?* Is pattern thinking truly as distinct from verbal thinking and visual thinking as verbal and visual thinking are from each other? Despite all the evidence over the centuries for thinking in patterns and despite all the recent research into thinking in patterns, people weren't talking about *pattern thinking* itself. Were they?

One Saturday evening, I went on a "surfin' safari." That's what I call it when I do a major, hours-long Internet search. I might start out with a goal in mind, but then I simply follow the trail through the jungle, from one piece of research to the next. On this occasion, my goal was to find scientific papers about a third type of thinking. Right away, of course, I found lots of papers about visual thinkers and verbal thinkers. And for nearly an hour, that's all I found. But then—there it was, in beautiful black and white: "Evidence for Two Types of Visualizers," read part of the title of one paper. Not two types of thinkers, verbal and visual, but two types of *visualizers.* Two types of visual thinking. And what were those two types? The title of another paper by the same lead author gave the answer: "Spatial versus Object Visualizers."

I quickly began searching for more papers by the same author, and I found a few. But when I went to the citation index — the list of other papers that cited these papers — the trail went cold. This small cluster of papers was *it:* a new branch of research, one that was finding empirical evidence to support my anecdotal hunch.

These papers and I were using different terms. What I called a *picture thinker,* these papers called an *object visualizer,* and what I called a *pattern thinker,* these papers called a *spatial visualizer.* But we were both saying the same thing: The old way of grouping all visual thinkers into one category was wrong.

This categorization had never been anything more than an assumption. This assumption was sensible in its own way, but not based on evidence. It was simplistic: *Visual thinkers are people whose thoughts rely on images.* Well, yes, they are. Jessy Park and I both see the world through images. Daniel Tammet and I both see the world through images. But we sure don't all see the world in the same way.

I called the author whose name had appeared (with an assortment of collaborators') on all these papers. Maria Kozhevnikov, a cognitive neuroscientist at the National University of Singapore, was a visiting professor of radiology at Harvard Medical School when I spoke to her. The conversation, I hoped, would provide some insights into the scientific rationale behind the need for a third category of thinking. I wasn't disappointed.

Kozhevnikov said that as a PhD candidate at the University of California, Santa Barbara, in the late 1990s, she had been looking at data from spatial tests — tests that ask you to manipulate images in space rather than just see them — when she noticed an odd artifact. Subjects who identified themselves as primarily verbal thinkers and those who identified themselves as primarily visual thinkers scored, on average, just about the same on spatial tests. That didn't seem right. You would expect people who think in pictures to be better at manipulating images than people who don't think in pictures.

She dug a little deeper into the data. And she noticed that while the visual thinkers' group average on the spatial tests was about the same as the verbal thinkers' group average, the visual thinkers' *individual* scores diverged along two extremes. Some scored very well. Some scored very poorly. They were all visual thinkers, yet some could easily manipulate objects in space, and some could not.

"It was clearly a bimodal distribution," she told me. "*Clearly.* It was so obvious from the statistical data that you had two types of people who report themselves as highly visual. One group had very high spatial ability, and the other group had very low. And I had the idea: Maybe the two groups are just different."

By then, researchers using new neuroimaging techniques had begun to establish the existence of two visual pathways in the brain. One is the dorsal (or upper) pathway, which processes information about the visual appearance of objects, such as their colors and details. The other is the ventral (or lower) pathway, which processes information about how objects relate to one another spatially. This view of the brain's division of labor soon became orthodoxy. In 2004, for example, researchers at a neuroimaging center at the Université de Caen and Université René-Descartes, in France, gathered the results from various PET studies conducted in their laboratory and found that higher activation in the dorsal pathway did seem to correspond to object imagery, and higher activation in the ventral pathway did seem to correspond to spatial imagery.

People obviously use both pathways, relying more on one or the other depending on the task. Kozhevnikov's challenge was to determine whether some people consistently use one pathway significantly more than the other, no matter what the task. Were some people dorsal — image — thinkers, and some people ventral — spatial — thinkers? As was the case when I considered this possibility, the more Kozhevnikov thought about it, the more sense it made. "Intuitively, you would expect this," she told me, "because visual art is so different from science" — two vocations that rely on visual thinking.

Kozhevnikov said her original paper presenting this hypothesis had been rejected by eight or nine educational journals. Editors said that maybe the visual thinkers who scored low on the spatial tests hadn't evaluated their own skills properly, or maybe they had abilities they didn't recognize, or maybe she wasn't taking into account gender differences, and so on. So she sent the paper to psychological journals, where it received a more welcoming reception.

In 2005 she published a paper that used behavioral data to argue for the existence of two types of visual thinkers — object and spatial. She and her colleagues then developed a self-report questionnaire to distinguish the two types of thinkers. She knew, however, that psychologists weren't going to be satisfied with only behavioral studies or self-reports. They would want evidence through neuroimaging — and in 2008, her team produced an fMRI study that showed that spatial and object visualizers do indeed use the dorsal and ventral pathways in different proportions.

Kozhevnikov's work is now widely accepted within her field; she receives "tons" of invitations to give talks on the subject, and the tests that she and her colleagues have designed over the years are frequently used in the United States, especially for personnel selection and assessments.

I asked her if I could take some of these tests myself, in order to better understand both my own thinking and thinking in general, and she generously assented.

The first test I took was called VVIQ, for Vividness of Visual Imagery Quotient. As the test's name suggests, its purpose is to identify how strongly a subject sees images in purely visual (as opposed to spatial) terms. It was divided into four sections, and for each section I had to imagine a different picture. One section directed me to imagine a relative or friend, another a rising sun, the third a shop I frequent, and the fourth a country scene involving trees, a mountain, and a lake. Each section consisted of four aspects of the image ("A rainbow appears," for instance, or "The color and shape of the trees") that I was to imagine and evaluate on a scale from 1 to 5 — from "No image at all (only 'knowing' that you are thinking of the object)" to "Perfectly clear and as vivid as normal vision."

Not surprisingly, I suppose, I gave nearly all the images in my mind 5s. When I read, "A rainbow appears," I immediately envisioned a rainbow that I had seen at a Chicago hotel a few days earlier; I'd actually gone outside to get a better look. When I read, "The front of a shop you often go to," I saw the King Soopers food market; I saw it from the front, I saw it as I walked in, I saw exactly where those little shopping baskets were.

The only images I didn't give 5s to were three of the four involving a friend. One instructed me to see the "*exact* contour of the face, head, shoulders, and body" (emphasis added), and boy, I saw them. And I saw them because I was asked to see *specific* details. I gave that image a 5. But in the next three images, I was asked to see more general aspects — one was "The different colors worn in *some* familiar clothes" (emphasis added) — and I had problems. The images I saw in those three questions I rated with 2s — "Vague and dim."

Still, when you added up the thirteen 5s and three 2s, my total VVIQ was 71 out of a maximum of 80. Kozhevnikov wrote back that this total was "VERY high," and "at the level of visual artists," whose mean was 70.19.

Next I tried the grain-resolution test. "Grain is density," the instructions explained, "defined roughly as: 'number of dots' per area (or volume)." For example, you can speak of the "graininess" of bumps on a raspberry or of spots on a leopard. Per unit of area, the raspberry has more bumps than the leopard has spots. Or think of goose bumps on your skin and then think of a spoonful of coffee beans. Which has a higher degree of graininess? If you said that tightly packed goose bumps have a greater graininess than larger and looser coffee beans, you're right. What about cottage cheese and cotton candy? If you think about the clumps in cottage cheese and the sugar granules in cotton candy, then you would see that cotton candy is grainier.

See is the key. The grain questionnaire, like VVIQ, is a test of object imagery, not spatial imagery. So for me, the test was a breeze. You ask me which is grainier, the briquettes in a heap of charcoal or the holes in a basketball net, and I *see* charcoal passing through a hole in a basketball net. You ask me which is grainier, a tennis racket or a bunch of grapes, and I *see* that I can't get an average-sized grape through a hole in a tennis racket's strings without squashing it.

The test consisted of twenty of these kinds of pairs, and I got seventeen out of twenty correct — though I filed a protest on one "incorrect" answer. Pavement or sponge? The answer key said pavement. I said sponge, but only because I didn't know what kind of pavement material the questionnaire meant! You tell me what you mean by pavement, and I'll tell you whether it's grainier than a sponge. Asphalt or concrete? When you lay down asphalt, you can see the aggregate — the base material that's made up of particles of various substances. Those lumps can be pretty big — bigger than the holes in a sponge. Even in concrete, the aggregate will show if the surface is worn down enough. In the days after I took this test, you'd better believe I went out and looked at pavement. I looked at all kinds of pavement. The steps in front of my building? They're floated concrete — the kind where the fine particles float to the surface. Okay, in that case, the answer key to the test was correct; floated concrete is grainier than a sponge. But the parking lot? I was right. Waiting at a light on Prospect Avenue, I opened my door and looked down. Right again. So you know what? I'm going to raise my score to 18.

What did I get wrong? Chicken skin and avocado skin. I've seen a lot of raw chickens in processing plants. The problem for me is that I don't cook, so I haven't had much experience with handling avocados. And the avocado slices I get in a restaurant on a salad are of course already peeled. But just to make sure that I actually did get this comparison wrong, I went to the supermarket and looked at an uncooked chicken and an avocado. Sure enough, chicken skin is grainier, the opposite of what I had answered.

Which leaves only shaving foam and sugar. Well, I hadn't used shaving foam in decades, so I had no idea what the answer was. I guessed shaving foam. Wrong. (But again, just to make sure, I went out and bought three types of shaving foam and conducted a comparison experiment in my kitchen. I can't imagine what the cashier thought.)

Still, my score of 17 was "VERY high," said Kozhevnikov. For visual artists, the mean is 11.75. For scientists and architects, she added, the mean is less than 9.

Now, that was pretty interesting to me. Twice I had scored in the same range as visual artists, and not in the same range as scientists. But I *am* a scientist. Then again, those were object imagery tests, and objects — pictures — are first nature to me. What would the spatial-relations tests show?

The first test I tried, each question began with a series of illustrations that showed a sheet of paper being folded. Let's say the first illustration showed a square piece of paper, then the next showed the paper being folded in half top to bottom, then a third showed the half sheet being folded in half again, left to right. The final illustration showed a pencil poking a hole in the half of a half of a sheet. The challenge was to imagine the sheet being unfolded back to its full size and then to compare the unfolded sheet in your mind with five illustrations on the page. Which illustration showing a sheet of paper

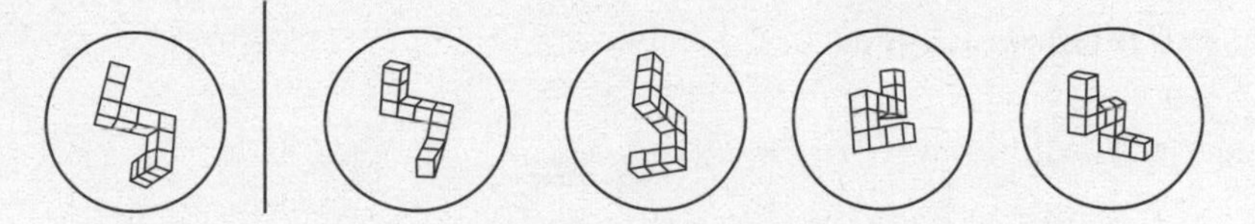

Take the 3-D object on the left and mentally rotate it, and it will match two of the illustrations on the right — but which two? The answer: the second and third. © *Houghton Mifflin Harcourt / Jay's Publishers Services; redrawn by permission from "Mental Rotation of Three-Dimensional Objects," by R. N. Shepard and J. Metzler, Science Magazine, February 19, 1971.*

with a hole or holes in it matched the one you were seeing in your mind?

This time I scored below average — four out of ten. Again, though, this score was consistent with visual artists', and it was the opposite of what scientists and architects scored.

Next I tried another spatial test. It showed a series of Lego-like blocks in various three-dimensional formations full of right angles. I do well at block design tests; I aced one recently while participating in a study at the University of Utah. *Aced it.* And in the allotted time. But that was a test that allowed me to touch the objects and manipulate them myself. The challenge with Kozhevnikov's test was to rotate each object *mentally* and then "see" which of the five accompanying illustrations it matched. I couldn't even *do* this test. My short-term memory is nearly nonexistent, so by the time I started rotating the object in mental space, I forgot what it looked like originally.

I have done a lot of thinking about the spatial relations test, I wrote to Kozhevnikov. *I can do well in certain types of visual spatial tests.* I explained that I could rotate a two-dimensional object — a flat drawing — in my mind. You show me the outline of Texas upside down and ask me what it is, and I won't hesitate: "That's Texas." But in my work, I actually don't have to rotate an object. *When I visualize a large cattle handling facility in my mind,* I wrote in my e-mail, *I move my mind's eye around it.*

Kozhevnikov considered this response, then sent back another

Spatial Orientation

Example:
Imagine you are standing at the **flower** and facing the **tree**.
Point to the **cat**.

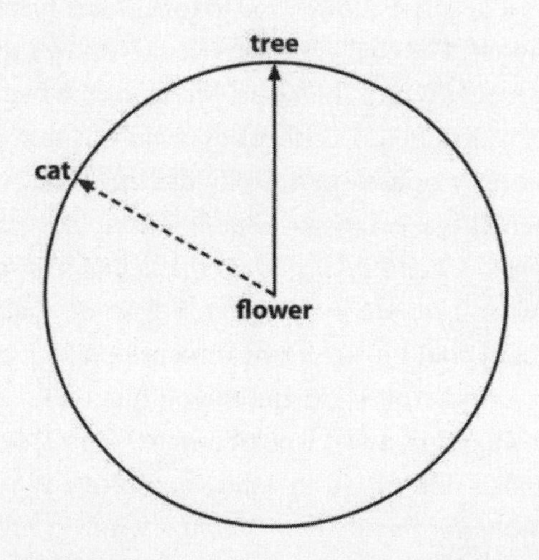

An example of the Perspective Taking / Spatial Orientation Test.

© Kozhevnikov & Hegarty (2001)

test and asked me to take it. Again it was a test of spatial abilities, but this time it didn't require me to imagine rotating an object. Instead, it asked me to change my perspective in relation to a landscape.

The test used one drawing over and over. It showed an assortment of various objects in random locations as if seen from above — a flower, a house, a stop sign, and so on. My job was to imagine myself (for instance) standing at the flower, facing the house, and pointing to the stop sign — then to render the angle of my pointing arm on a circular graph, with myself at the center. Now, I know I'm good at judging angles. I can look at a ramp in a cattle facility and say, "That's at a twenty-degree angle," and I'll be right. Guaranteed. But this test required me to imagine myself hovering above the scene and see the angles from the perspective of a person standing below. Let me tell you, that's not the same as standing on the ground and looking out of my own two eyes. Anyway, at least I could complete this test. Not that it mattered: I scored zero.

These results made no sense to me whatsoever. When I taught myself how to draw blueprints, years and years ago, I walked around an entire Swift meatpacking plant matching every line on the plant's original architect's blueprints with its corresponding real structure. For example, a big circle on the blueprints was the water tower, and a little square was a concrete column that held up the roof. This exercise taught me how to relate the abstract lines on the blueprints to the actual structures. When I do remodeling jobs and I have to figure out how to fit new equipment into an existing place where some parts have to be torn out, I spend fifteen to twenty minutes just looking at the site, until I feel I've fully downloaded all the visual details into my memory. When I test-run equipment in my mind, I can move myself around the image. I can fly over it, walk through it, walk around it. I can see it from the point of view of a helicopter looking down at the whole facility, and I can see it from the point of view of an animal walking along at ground level.

When I'm consulting on or designing a project that doesn't yet exist, I scroll through my memory bank looking for similar images. To demonstrate how this process works for me, I asked Richard, my collaborator on this book, to suggest something for me to design in my mind. He said, "A fence."

"A fence?" I said. "What kind of fence? For what? A cattle fence? A fence alongside a highway? A privacy fence at a house? Barbed-wire fences? Picket fences? Wooden-plank fences, plastic fences, fake white-plank fences? Wrought-iron fences? Pipe corrals? Solid sides on cattle handling facilities?" These were all coming up as pictures in my mind. "There is no *fence*."

Needless to say, Richard is not autistic.

He tried again. He said that he'd recently seen on television a design for a bridge between Hong Kong and China. In Hong Kong, cars drive on the left side of the road (because it's a former British colony), and in mainland China, cars drive on the right side. How would I design such a bridge?

"I'm seeing roadways crossing," I said. "I see my little brother's

The switchover bridge does what it says.

© NL Architects; Flipper Bridge, switching lanes between mainland China and Hong Kong

slot-car track. I see a woven hanging basket with a flower pot in it. Now I'm seeing freeway ramps — specific freeway ramps. I see roads. Okay," I said, ready to give my answer. "It would have to have an underpass and an overpass, and the roads would cross and switch sides."

Richard told me to Google *flipper bridge*. The image that came up on my computer screen was the one I'd seen in my mind.

Sometimes when I'm consulting, company executives will take me to a meeting room and show me the specifications on the project, and I'll just sit there and run the "movie" in my head. I'll see exactly how the design is going to play itself out, and I'll say something like, "That's not going to work. It's going to jerk on the chains too hard and rip 'em right out of the ceiling."

I used this technique on a couple of exercises in a paper by Kozhevnikov and collaborators. The topic of the paper was how different kinds of minds handled problems in physics. One exercise (see illustration, below) asked you to imagine a hockey puck traveling in

A hockey puck is traveling in a straight line, a to b. When it reaches b, it receives a heavy kick in the direction of the heavy print arrow. Which of the paths below will it follow? © *David Hestenes*

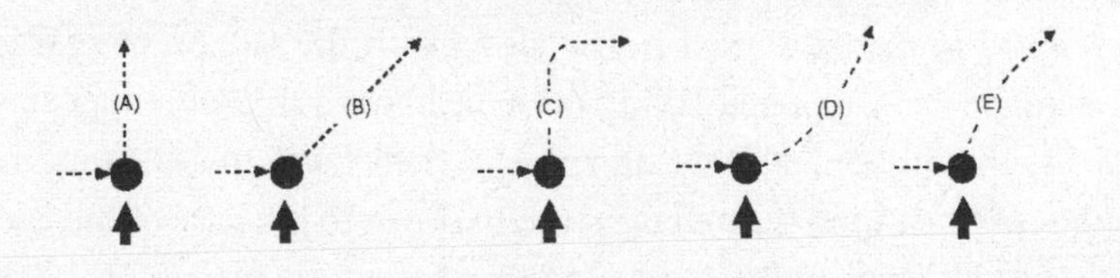

The answer is (B) — a straight line angling away from the kick.
© *David Hestenes*

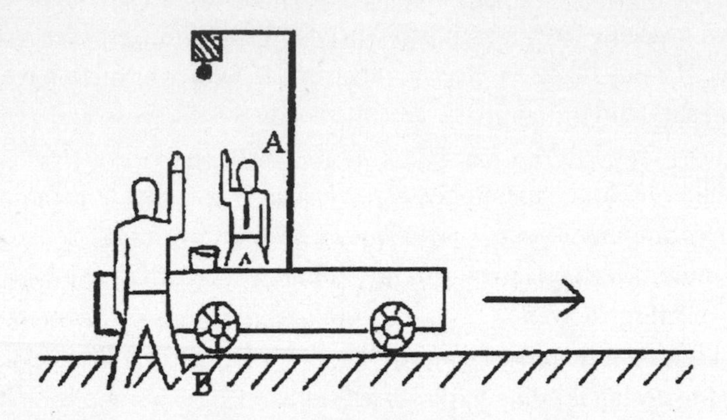

© Maria Kozhevnikov

a straight line until it received a single kick from a foot striking it at a right angle to its path. Where will the puck go? The answer, I immediately saw, was a straight line angling away from the kick. And I saw it because I could run that movie.

Same thing with this problem (see illustration, above): A ball sits at the top of a mast on a cart that is traveling on a straight road. If the ball drops from the top of the mast to the base of the cart, how will the trajectory look to someone riding along with the ball on the cart? It will look like the ball is moving straight down. How will it look to an observer watching the cart from the side of the road? It will look like the ball is moving forward as it travels along with the cart. How did I know? Because I could run the movie in my mind.

When I imagined the ball falling from the mast to the base of the cart, and I imagined myself riding along with the ball on the cart, I immediately saw a pencil falling off a dashboard in a moving car—and I saw that it fell straight down. And then I saw myself standing on the cart watching the ball fall straight down to the base of the cart.

I wrote to Kozhevnikov and confessed my confusion about the results of the spatial tests. *When I do photography,* I wrote, *I can determine from the ground the best place to stand on a roof to get the best shot.* I've done just that with professional television and movie crews. "You want the perfect shot of the cattle?" I'll ask them. "Go up on the corner of that roof over there and face the feedlot." How can I not be a spatial thinker?

Kozhevnikov wrote back that in imagining the scene from the roof, I'm not manipulating an object in space. I'm manipulating *me* in space. I'm visualizing an object from a new perspective, but I'm still visualizing an *object.* I'm still *thinking in pictures.* When I'm drawing a blueprint, remodeling a plant, or designing a project, my thinking starts with an image of an object. Even the movies in my head start with a still image.

Which is why I scored the way I did on the tests. On the object imagery tests, I had scored high — as high as visual artists, and even higher. On the spatial imagery tests, I had scored low — as low as visual artists, and even lower. I am a visual thinker, and in both sets of tests my scores were remarkably similar to those of visual artists. But how to account for the fact that I'm a scientist, yet where I scored high, scientists scored low, and vice versa?

Richard took the tests too. He scored perfectly on the spatial tests — the paper folding, the mental rotation, the stand-at-the-flower-and-face-the-house-and-point-at-the-stop-sign. But the grain test presented problems for him; he got eleven right out of twenty. Not bad, but not in the category of pulling up images of two objects and comparing them, the way I do. Because he's a writer, he identifies himself as a verbal thinker. The visual tests showed that he also has superior spatial abilities, similar to a scientist's. Is it any wonder, then, that even though he's not a scientist, he specializes in *writing* about *science?*

The correlation between how the tests predicted he would think and how he actually does think was simple, direct, clear. Yet the same tests told me I was the kind of thinker that I knew for a fact I wasn't. Why?

The answer was autism. I found an exercise in one of Kozhevnikov's papers that showed two abstract paintings. The first consisted of big swooshing splashes of color; the whole impression of the painting was dynamic. The second showed various sorts of geometrical shapes; the impression was static. When I looked at the splashy, dynamic painting, I immediately saw a picture of a fighter jet that I had just seen in a book I was reading. When I looked at the static painting, I immediately saw Mother's sewing basket.

"What kind of feelings does that bring up in you?" Richard said when we were discussing these paintings.

"Feelings?"

"What kind of emotional response do you have when you see your mother's sewing basket?"

"None," I said. "I see Mother's sewing basket when I look at that painting because it looks like Mother's sewing basket to me. I also see a salad I ate last week at the restaurant where I sometimes like to have lunch. They put Wheat Chex on their salads instead of croutons. I look at that painting, I see a picture in my mind of Mother's sewing basket and I see another picture in my mind of Wheat Chex on a salad."

Still, I understood what Richard was saying. Another person might have an emotional attachment to his mother's sewing basket, an object he remembers fondly from his childhood. And in fact, Kozhevnikov's research showed that in describing the two paintings, artists used *emotional* terms — *crash, breakthrough, extreme tension.*

I *see* like an artist, I realized, but I don't *feel* like an artist.

Instead, my emotions work like a scientist's. When scientists described the paintings, they used *unemotional* words — *squares, stains, crystals, sharp edges,* and *swatches.* I'm not saying that scientists and engineers don't feel emotions; I'm sure most scientists and engineers would feel some sort of sentiment about their mothers' sewing baskets. But the scientists in this study didn't see a mother's sewing basket, or any other object. They saw geometrical shapes. They saw what was *literally* there, and what was literally there wasn't the kind of image that would elicit an emotional response. Artists, on the other hand, saw what was *figuratively* there, and what was figuratively there was indeed the kind of image that would elicit an emotional response. I saw what was figuratively there too — only those images did *not* produce an emotional response in me.

Like Michelle Dawson, who described autistic traits not as positive or negative but as accurate, I don't attach an emotional response to concrete objects. So I am able to handle them objectively — literally *as objects,* and only as objects. I can't manipulate them in space. I can't subject them to spatial reasoning. But I can sure design a cattle chute that works.

That's why there are certain design mistakes that I would never make, even though some engineers make them. Some engineers use spatial visual thinking, but I use object visual thinking, so I'm able to *see* a catastrophe before it happens. Airbags in cars killed many children because the engineers blindly followed a bad specification — that in an accident, the bag must be able to protect an adult man who was not wearing a seat belt. If I had watched videos of the crash-dummy tests, I could have easily seen that babies would not survive the airbag's impact. During the Japanese tsunami catastrophe of 2011, the Fukushima nuclear power plants melted down because the tidal wave that came over the seawall flooded not only the main generator but its backup. And where was the backup located? In the basement — the basement of a nuclear power plant that is located next to the sea. As I read many descriptions of the accident, I could *see* the water flowing into the plant, and I could *see* the emergency generators disappearing under the water. (This is partly what I do as a consultant: I see accidents before they happen.)

So my test results were consistent after all. The correlation between how the tests predicted I would think and how I actually do think was simple, direct, clear — once I factored autism into the equation:

high object imagery plus autism equals scientific mind, at least in my case.

Now that I'd satisfied myself that the three-kinds-of-minds hypothesis makes sense, I had to ask: Could it be useful in helping the autistic brain?

8 From the Margins to the Mainstream

REMEMBER JACK? He was the boy who could ski better after three lessons than I could after three years, because I was the one with the cerebellum that's 20 percent smaller than normal. But you know what I *could* do? Draw. Design.

And so sometimes while Jack was getting in a lot of ski practice, I stayed at the top of the slope and got to work—my kind of work. I refinished the ski-tow house. I installed knotty-pine boards and stained them; I added white trim; and I made a nice sign showing the insignia of my school. I took an ugly plywood shack, and, because of who I am, I made it into a thing of grace—a grace that my physical movements, also because of who I am, would never be able to match.

That experience was an early lesson in how I can play to my strengths. I didn't think of myself as a picture thinker back then, of course. But I knew that drawing was not only what I could do, it was what I could do best. And so I did it. I took what nature gave me, and I nurtured the heck out of it.

In recent years, the relationship between nature and nurture has been getting a lot of attention in the popular press. In particular, the 10,000-hour rule seems to have captured the public imagination. *New Yorker* writer Malcolm Gladwell didn't invent the rule, but he did popularize it through his best-selling book *Outliers.* The principle actually dates to a 1993 study, though in that paper the authors called it the 10-year rule. Whatever name it goes under, the rule essentially says that in order to become an expert in any field, you need to work for at least *x* amount of time.

I don't know what all the fuss was about. After all, the old joke goes, "How do I get to Carnegie Hall?" "Practice, practice, practice," not "How do I get to Carnegie Hall?" "Be born with talent, then do nothing." But I guess a big round number brings the equation to life or makes a formula for success sound scientific in a way that simply saying "Practice, practice, practice" doesn't. Still, that interpretation of the rule seems reasonable to me. Talent plus ten thousand hours of work equals success? Talent plus ten years of work equals success? Sure!

But that's not how the rule often gets interpreted. Consider an article about the 10,000-hour rule in *Fortune.* It was published in 2006, but it's still widely posted all over the Internet. The article opens with the example of Warren Buffett, one of the wealthiest people in the world. "As Buffett told *Fortune* not long ago, he was 'wired at birth to allocate capital.' . . . Well, folks, it's not so simple. For one thing, you do not possess a natural gift for a certain job, because targeted natural gifts don't exist. (Sorry, Warren.)"

Maybe the issue here was the word *targeted.* Was Warren Buffett born to be a CEO *specifically?* Was he born to run a behemoth corporation like Berkshire Hathaway rather than, say, to work as a day trader? No. But was he born with a brain for business—a brain that would lend itself to number-crunching and risk-taking and opportunity-identifying and all the other skills that go into becoming the leading investor of his generation? I say yes.

Certainly Buffett put in his ten thousand hours or ten years of work. He bought his first shares of stock at the age of eleven, founded a successful pinball-machine business with a friend at the age of fifteen, and before he graduated high school, he was wealthy enough to buy a farm.

But this is not the career trajectory of someone who's interested in business and is putting in his ten thousand hours. This is the career trajectory of someone who lives to do business. You might say it's the path of someone who was born to do business. You might even say it's the path of someone who was wired for business at birth.

By putting such an emphasis on practice, practice, practice *at the expense of* natural gifts, the *Fortune* interpretation of the 10,000-hour rule does a tremendous disservice to the naturally gifted.

But wait. It gets worse. Some interpretations of the 10,000-hour rule leave talent out of the equation altogether.

Here is the description of the 10,000-hour rule on a website called Squidoo (a worldwide community that, like Wikipedia, allows users to create brief entries on popular topics): "If you want to become an expert in your field, be that art, sport or business—you can. Contrary to popular belief, it's not always innate genius or talent that will make you a success, it's the hours that you put in, which means that ANYONE can do it."

Well, no. Not *everyone* can do it. Let's go to Gladwell's example of Bill Gates. In the late 1960s, when Gates was still in high school, he had access to a Teletype terminal, and his math teacher excused him from class so that he could write code. Computer code became something of an obsession with Gates, and ten thousand hours later—well, you know the story.

Now let me tell you the other side of that story. In the late 1960s, when I was a student at Franklin Pierce College, I had access to the same terminal as Gates—the *exact same* Teletype terminal. The school's computer system tapped into the University of New Hampshire's mainframe. So I had as much access as I wanted, and I had as much firepower as I wanted, and it was all free. And you'd better believe I wanted to spend as much time as possible on that computer.

I love that sort of stuff; I love to see how new technology works. The computer was called Rax, so when I turned on the computer, a message would type out on paper: *Rax says hello. Please sign in.* And I would eagerly sign in.

And that was it. I could do that much — but that was all.

I was hopeless. My brain simply doesn't work in a way that allows me to write code. So saying that if I'd spent ten thousand hours talking to Rax, I would be a successful computer programmer, because *anyone* can be a successful computer programmer, is crazy.

I say:

Talent + 10,000 hours of work = Success

Or to put it another way:

Nature + nurture = Success

Squidoo says:

10,000 hours of work = Success

Or to put it another way:

Nurture = Success

Stated so baldly, this interpretation of the 10,000-hour rule looks ridiculous. Like *Fortune*'s analysis of Warren Buffett's success, Squidoo's interpretation does an injustice to the naturally gifted. But it also does a tremendous disservice to the naturally *ungifted.* It raises hopes to an unrealistic level. All the hard work in the world won't overcome a brain-based deficit (like a cerebellum that's 20 percent smaller than normal).

Neuroanatomy isn't destiny. Neither is genetics. They don't define who you will be. But they do define who you might be. They define who you can be. So what I want to do here is focus on how the autistic brain can build up areas of real strength — how we can actually change the brain to help it do whatever it does best.

The idea of plasticity in the brain — that your brain can create new connections throughout your whole life, not just in childhood — is still quite new, and like so many new ideas about the brain, we owe our awareness of it to neuroimaging. Until the late 1990s, scientists tended to think that the brain remained essentially the same, or even deteriorated, over time. One particularly compelling finding that helped overturn this view was a 2000 study of London taxi drivers. In order to qualify for a license, a London taxi driver has to learn what's known as the Knowledge — the location of every nook in the city, and the quickest way to get there. Specifically, he needs to memorize the names and locations of the twenty-five thousand streets that radiate from central London, a task that takes the average person two to four years. And the prospective cabdriver needs to demonstrate this knowledge in a series of tests taken over the course of several months. These tests consist of one-on-one interviews with inspectors who name a point of departure and a point of arrival; the applicant's job is to describe how to make that trip, turn by turn.

A study led by Eleanor Maguire, a British neuroscientist, looked at MRIs of the hippocampi of sixteen licensed London cabbies. The hippocampus is believed to house three types of cells that help us navigate: place cells, which recognize landmarks; head-direction cells, which tell you which way you're facing; and grid cells, which tell you where you are in relation to where you've been. What Maguire found was that the hippocampi of drivers who had mastered the Knowledge were larger than those of control subjects. What's more, the longer a driver had been on the job, the larger the hippocampus.

And what happens when a driver leaves the job? In a follow-up study, Maguire found that the hippocampus returns to normal size.

"The brain behaves like a muscle," Maguire said. "Use brain regions and they grow."

But if you don't use a brain region, it won't necessarily wither. Neuroscientists have been intrigued by a case in India: A man who had been nearly blind since birth had his vision restored. SK (as he was known) had congenital aphakia, a condition in which the eyeball develops without a lens. He had 20/900 vision — that is, he could make out at twenty feet what people with regular vision could make out at nine hundred feet. For SK, the world was a shadowy landscape. When he was twenty-nine years old, some visiting doctors gave him a pair of glasses. His visual acuity improved to 20/120, but his doctors didn't know if he would ever be able to make sense of what he saw. For example, he could see patches of black and white, but until those patches moved, he had no idea they were parts of a cow. Initially, his visual skills were rudimentary. He could recognize some basic two-dimensional objects, but nothing beyond that.

And for some time, that's where the quality of his vision remained. His lack of progress was not surprising, at least according to the neurological theory that the brain has a window of opportunity in which to develop vision. Miss that window — which comes very early in life — and it shuts forever.

Yet about eighteen months after receiving his glasses, SK could recognize some complex objects. He could distinguish colors and levels of brightness that had previously eluded him. He didn't need the cow to move to know it was a cow.

He could see.

What had changed wasn't his vision. It was the way his brain processed the images. His eyesight was still 20/120, but now he could interpret images in a new way. His brain had needed time to adapt.

Because of SK, researchers have had to jettison a lot of their ideas about how vision develops in the brain. Now they'll have to see if they can help blind children who are older than eight — the previous standard cutoff point. They'll have to see what neuroimaging reveals. As one neuroscientist marveled, "People can learn to use the vision they have."

Not only can dormant areas of the brain "come to life" and do what they were always supposed to do, but those areas can get repurposed and do what they *aren't* supposed to do.

Researchers at the Massachusetts Eye and Ear Infirmary have developed a method to research the brain activity of people who have been blind since birth. It works like a video game. Players have to navigate through a building in a search for diamonds. But the game doesn't use images. It uses sounds.

Players figure out where they are and where danger lurks by *listening* to their environment in 3-D sound instead of looking at it. Footsteps echo. The sound of a knock indicates the location of a door. A ping means the player has bumped a piece of furniture. The diamonds make a twinkling noise that grows louder as the player approaches.

The layout of the labyrinth actually corresponds to an administrative building next to the research lab — a place the players wouldn't have visited. But when they finish the game and go into the building, they know their way around immediately. When a similar experiment was tried on both blind and sighted children in Santiago, Chile (where the research originated), the sighted subjects playing the game didn't even realize that what they were supposed to be "inside" were corridors in a building.

Over the years, scientists have used PET scans and fMRI scanners and MRI machines to study the visual cortex (which covers 30 to 40 percent of the brain's cortical surface) of subjects who have been blind since birth. They have found that even though the blind person's visual cortex had never received any visual stimulation, it was nonetheless being used. In effect, it had been repurposed to perform the blind equivalent of visual tasks such as reading (Braille), localizing sounds, interpreting body language, and so on.

These results were consistent with what the Massachusetts researchers found when they looked at the brain activity of blind-since-birth players of the "video" game. They also found that when a sighted subject needed to make strategic decisions, he used the hippocampus, the brain's memory center. But the blind subject used his *visual* cortex.

I witnessed some similarly remarkable abilities in the behavior of my blind roommate in high school. I called her a "cane master." She didn't want a guide dog leading her. She wanted to learn how to guide herself. And boy, did she ever. She needed to be walked through a new environment only once, and then she knew her way. Outside our dorm was a busy intersection; she navigated it as well as any sighted person. Now I can look back at what she was doing and have at least a little insight into how she was doing it. In a way, she really *was* seeing her environment. Maybe she wasn't using actual images, but her *visual* cortex was allowing her to build a vivid, knowable, and navigable world.

A change in one part of the brain can also apparently lead to changes in other parts of the brain. I helped a dyslexic graduate student of mine overcome some of her visual problems through the use of tinted eyeglasses. They did the job — her eyesight got better, and she graduated to lighter and lighter tints until finally she didn't need the glasses at all. But the correction to her vision helped correct other problems that you might think were unrelated. The organization of her writing improved. Suddenly she was expressing herself on paper with greater ease and clarity.

I don't know how my own brain might have changed over the years, but I do know that as my career has shifted, so have my abilities. I haven't been doing drawings for more than ten years now, partly because of changes in the industry. The fax machine was the ruination of good architectural drawings. Clients would say to me, "Oh, just shove it in the fax," and then they'd use the fax as their blueprint. I lost the motivation to make a really nice drawing. But at the same time, my professional priorities were changing. I was becoming a lot busier giving lectures, and many people have told me that my speaking style became more and more natural. That was hard work. I knew I had to train myself to be someone I wasn't naturally, and what is training yourself at a new skill but "rewiring" your brain?

This generation is fortunate in an important way. They're the tablet generation — the touchscreen, create-anything generation. I've already talked about how these devices are an improvement over previous computers because the keyboard is right on the screen; autistic viewers don't have to move their eyes to see the result of their typing. But tablets also have other advantages for the autistic population.

First, they're cool. A tablet is not something that labels you as handicapped to the rest of the world. Tablets are things that normal people carry around.

Second, they're relatively inexpensive. They're even less expensive than high-end personal communication devices traditionally used in autism classrooms.

And the number of apps seems limitless. Instead of a device that performs a few functions, a tablet taps into a world of educational opportunities. You have to be careful, of course. I saw an educational app that visually was quite cute — it featured Dr. Seuss characters — but its approach was inconsistent. If you touched the image of a ball, the tablet said, "Ball." But if you touched the bicycle, it said, "Play," and if you touched the wall, it said, "House." Those words are too abstract. It needs to say, "Bicycle," and it needs to say, "Wall." But the better programs and apps say what they mean, and they can be invaluable in helping nonverbals communicate.

These days you can get a whole education online. Numerous websites and high-tech tools that offer amazing opportunities have cropped up. The names and aims of these sites will undoubtedly change over the years, but at the moment here are some of my favorite educational accessories that are perfect for some autistic brains.

- Free videos. Khan Academy offers hundreds if not thousands of educational videos and interactive graphics in dozens of categories. You're a pattern thinker who wants to know more about computer programming? Try the code-writing-for-animation category. You're a picture thinker? Browse the hundreds of art history videos that cover historical movements, geographical specialties, and individual artists and artworks.

- Semester-long courses. Coursera offers free courses from more than thirty universities. And the courses are changing all the time. Your kid is a science geek who's interested in the universe? You're in luck. A professor from Duke University is teaching a nine-week Introduction to Astronomy course, three hours of video instruction per week. You're a word-fact thinker who wants to write poetry? Learn from the masters with Modern and Contemporary American Poetry, a ten-week course taught by a University of Pennsylvania instructor. Udacity is another gateway to free courses, though ones with a more mathematical emphasis.
- Check out the universities themselves. I just typed *Stanford* and *free courses* into my browser, and up came a list of sixteen courses for the fall semester, including Cryptography and A Crash Course on Creativity. In 2012, Harvard, MIT, and the University of California at Berkeley created a nonprofit partnership in free courses called edX.
- 3-D drawing tools. They're free, they're downloadable, and they range in complexity. My personal favorite is probably SketchUp.
- Desktop 3-D printers. The programs—like SketchUp—are free, and the printers are dropping in price. Yes, they're expensive at the very moment I'm writing this sentence—about $2,500 for a low-end but perfectly serviceable model. But at the rate technology changes, that price has probably dropped to $2,400 in the time it took me to write *this* sentence.

I'm certainly not saying we should lose sight of the need to work on deficits. But as we've seen, the focus on deficits is so intense and so automatic that people lose sight of the strengths. Just yesterday I spoke to the director of a school for autistic children, and she mentioned that the school tries to match a student's strengths with internship or employment opportunities in the neighborhood. But when I asked her how they identified the strengths, she immediately started talking about how they helped students overcome social deficits. If even the experts can't stop thinking about *what's wrong* instead of *what could be better*, how can anyone expect the families who are dealing with autism on a daily basis to think any differently?

I'm concerned when ten-year-olds introduce themselves to me and all they want to talk about is "my Asperger's" or "my autism." I'd rather hear about "my science project" or "my history book" or "what I want to be when I grow up." I want to hear about their interests, their strengths, their hopes. I want them to have the same advantages and opportunities in education and the marketplace that I did.

I find the same inability to think about children's strengths in their parents. I'll say, "What does your kid like? What is your kid good at?" and I can see the confusion in their faces. *Like? Good at? My Timmy?*

I have a routine I follow in these cases. What's your child's favorite subject? Does he have any hobbies? Does she have anything she's done—artwork, crafts, *anything*—that she can show me? Sometimes it takes a while before parents realize that their kid actually has a talent or an interest. Two parents came up to me recently and said they were concerned because they knew their son wouldn't be able to handle the family business, a ranch. What would become of him, since that was the only world he'd ever known? Well, yes, it might be the only world he'd ever known, but the kid wasn't nonverbal. Their kid could function. So what part of that world interested him? Fifteen minutes later, they finally said that their son liked fishing.

"So maybe he can be a fishing guide," I said.

I could almost see the light bulbs popping to life above their heads. They now had a way to rethink the problem. Instead of thinking only about accommodating their son's deficits, they could think about his interests, his abilities, his strengths.

For me, autism is secondary. My primary identity is as an expert on livestock — a professor, a scientist, a consultant. To keep that part of my identity intact, I regularly block out chunks of the calendar for "cattle time." The month of June? That's cattle time. The first part of January? That's cattle time. I don't take speaking engagements during those periods. Autism is certainly part of who I am, but I won't allow it to define me.

The same is true of all the undiagnosed Asperger's cases in Silicon Valley. Being on the spectrum isn't what defines them. Their jobs define them. (That's why I call them Happy Aspies.)

Some people, of course, will never have that opportunity. Their difficulties are too severe for them to cope without constant care, no matter how hard we try.

But what about those who can cope? And what about those who can't cope but who *can* lead more productive lives if we can identify and cultivate their strengths? How can we turn the plasticity of the brain to our advantage?

Okay, let's take it one step at a time. First things first: How do we identify strengths?

One way is to apply the three-ways-of-thinking model that I discussed earlier: picture thinker, pattern thinker, word-fact thinker. That model, I believe, can help fundamentally change education and employment opportunities for persons with autism.

Education

When I give lectures in Silicon Valley, I see a lot of people who are solidly on the autistic spectrum, and then when I travel around the country and speak at schools, I see a lot of similar kids who will never get the chance to work in Silicon Valley. Why? Because their schools are trying to treat the kids like they're all the same.

Putting kids who are on the spectrum in the same classroom as their nonautistic peers and treating them the same way is a mistake. For elementary school children, being in the same classroom with their normal peers is good for socialization. The teacher can bring in higher level work in subjects the child excels at. But if a school treats everyone the same, guess what: The person who's not the same is going to stand alone. That person will be marginalized in the classroom. And once that happens, it won't be long before that student is marginalized for good — sent to a separate classroom or even a separate school. And suddenly the Asperger's kid might find himself in the same program as a bunch of nonverbal kids.

If you've read some of my other books or seen the HBO movie of my life, then you know the tremendous debt I owe Mr. Carlock, my high-school science teacher. He changed my life in many ways by identifying my strengths — mechanics and engineering — and helping me explore them. He ran the model rocket club, which I loved. He got me interested in all sorts of electronics experiments.

But in one crucial respect, his thinking probably held me back.

When Mr. Carlock saw that I couldn't do algebra — just could not do it — he redoubled his efforts to make me learn it. He didn't understand that my brain doesn't work in the abstract, symbolic way that solving for *x* requires. Mr. Carlock wasn't someone who liked to give up on a student, and I'm sure he thought that by pushing me hard on algebra, he was helping me. But what he could have done instead is recognize my limitation in that area and play to my strength in another area.

My engineering talent should have been a clue. Engineering isn't abstract; it's concrete. It's about shapes. It's about angles. It's about *geometry.*

But no. The standard high-school curriculum says algebra comes before geometry, and geometry comes before trigonometry, and trigonometry comes before calculus, and that's that. Never mind that you don't need to know how to do algebra in order to study geometry. Mr. Carlock, like a lot of educators, was stuck in a curriculum rut and didn't even realize it.

When I bring up this anecdote at my public appearances, I ask if anybody else had a similar experience. Always, four or five hands will go up. If an autistic fourteen-year-old can't handle algebra because it's too abstract, you don't say, "Do algebra anyway." You try moving him to geometry! If another kid can't handle algebra or geometry or any other kind of math, you don't say, "You have to do math before you can do anything else." Instead, try turning her loose in the lab! If a kid can't handle handwriting, let him type. If a kid like me invents something like the squeeze machine, you don't say, "That kid should be like other students" and then destroy the machine; you say, "That kid isn't like the other students, and that's a fact." The educator's job—the role of education in society—is to ask, "Well, what *is* she like?" Instead of ignoring deficits, you have to accommodate them.

Just the other day, I heard from a mother that her daughter couldn't handle the noise of the lunchroom, so the principal let her eat in the faculty lounge. The mother was upset that the principal had segregated her daughter. But I told her that no, this is a perfect solution to her daughter's problem. The principal was sensitive enough to recognize what her daughter could and could not handle and to find a creative way to accommodate her deficit.

But if you really want to prepare kids to participate in the mainstream of life, then you have to do more than accommodate their deficits. You have to figure out ways to exploit their strengths.

How do you do that? How do you recognize a strength when you see it? This is where the three ways of thinking—picture, pattern, and word-fact—come in handy.

I recently had a conversation with a parent whose fourth-grader was exceptional at art, but the school wanted to discourage him because his extreme devotion to drawing was "not normal." *He's a picture thinker!* I thought. *Work with it!* Don't try to make him into something he's not or, worse, into something he can't be. What you want to do instead is encourage his art—but broaden what his art encompasses. If he's drawing pictures of racecars all the time, ask him to draw the racetrack too. Then ask him to draw the streets and buildings around the racetrack. If he can do that, then you've taken his weakness (obsessional thinking about an object) and turned it into a strength (a way to understand the relationship between something as simple as a racecar and the rest of society).

Unless the child is a true prodigy or a savant, you're not going to be able to tell what kind of thinker she is at the age of two. In my experience, evidence of a predisposition toward picture, pattern, or word-fact thinking doesn't emerge until second, third, fourth grade.

Kids who are *picture thinkers* are the ones who like hands-on activities. They like building with Legos, or painting, or cooking, or woodworking, or sewing. They might not be good at algebra or other forms of math, but that's fine. You can work math into their hands-on activities. If the kid is into cooking, for instance, you can work fractions into the lesson—half a cup of this, a quarter cup of that. You can teach geometric shapes through origami. I would have understood trigonometry from building model bridges and destructively testing them—trying spans of different lengths, putting them at different angles, and seeing how much weight I needed to break the bridge. (Remember, concrete is just grown-up cardboard.)

Unfortunately, today's educational system is letting these kids down. It's phasing out hands-on classes, like shop—precisely the kind of class where geeky kids can feel at home and let their imaginations roam. I was at a processing plant recently to see a demonstration of robots that do some of the difficult, dangerous jobs. I asked who programmed the robots, and I was told it was done by five people from China and India. So I asked why they didn't use people from the United States. Because, I was told, our educational system doesn't produce bright young minds with the right combination of electrical engineering and computer engineering.

It's as if the word-fact thinkers have taken over the educational system. I know that the economy can be difficult and money is always tight, but we're talking about the future of a generation—or more.

Like picture thinkers, *pattern thinkers* tend to love Legos and other construction toys, but in a different way. Picture thinkers want to create objects that match what they see in their imagination, whereas pattern thinkers think about the ways the parts of the object fit together.

I was horrible at understanding word problems in physics. I couldn't even figure out how to put the problems together, because they placed too heavy a burden on my working memory. But if I had to solve a physics problem now, I would know what to do. I'd get five textbooks, sit down with a tutor and a spreadsheet, identify specific examples of problems that use one formula and specific examples of problems that use another formula, and eventually I would recognize the patterns in the problems.

A pattern thinker, however, would see the patterns a lot earlier. That's what makes pattern thinkers good at math and music: They *get* the form behind the function.

Many pattern thinkers, though not all, gravitate toward music. Pattern thinkers might find reading a challenge, but they'll be miles ahead of their classmates in algebra, as well as in geometry and trigonometry. It's important for schools to let them work at math at their own pace. If they're ready for a math text that's two grades away, give them that math text. Jacob Barnett, at the time a preteen autistic living in suburban Indianapolis, was so bored in grade-school math class that he started to hate math. Finally, out of frustration, he sat down with a bunch of textbooks and taught himself the entire high-school math curriculum in two weeks. Then he went to college—at the age of twelve.

It's also important for schools to let math whizzes do math in their own style. If they can do math in their heads, don't tell them, "You have to show your work." Let them do it in their heads. (Though you have to make sure that they're not cheating somehow. A simple electronic devices–free test in an empty classroom will answer that question.)

You'll know who these *word-fact thinkers* are because they'll tell you. They'll recite all the dialogue from a movie. They'll rattle off endless statistics about baseball. They'll calmly recall all the important dates in the history of the Iberian Peninsula. Their math skills will be only average, they won't bother with the Legos and building blocks, and they won't be all that interested in drawing. In fact, there might well be little point in forcing them to sit through art class.

One way to help this kind of thinker learn to engage with the world is to encourage writing. Give them assignments. Let them post on the Internet. (Word-fact thinkers tend to have strong opinions, in my experience, so just make sure to monitor their Internet use for safety — which is good advice when supervising any child.)

Employment

About fifty thousand people diagnosed with ASD turn eighteen every year in the United States alone. That's a little late to be thinking about adulthood. I tell parents that by the time their ASD kids are eleven or twelve, the parents should be thinking about what the kids are going to do when they grow up. Nobody needs to make a final decision at that point, but the parents should start considering the possibilities so that they have time to help prepare the child.

I've said it before, but I can't say it enough: Parents and caregivers need to get the kids out into the world, because kids are not going to get interested in things they don't come into contact with. This point might seem obvious, but I am constantly meeting individuals with Asperger's or high-functioning autistics who are graduating from high school and college with no job skills. Their parents have let them fall into a routine that never varies and that offers no new experiences. I didn't become interested in cattle until I went to my aunt's ranch. A high-school experimental-psychology class that featured lots of fascinating optical illusions stimulated my interest in both psychology and cattle behavior. The world is full of fascinating and potentially life-altering things, but kids aren't going to adopt them if they don't know about them. (Even autistic people with severe problems need to see the world. See chapter 4 for desensitizing tips.)

Of course, an ASD kid doesn't have to go visit an aunt in another state for inspiration. Sticking close to home will do just fine too. Not *at* home, but close to home. It's essential for him or her to get outside the house and accept responsibility for tasks that other people want done—and that need to be done on *their* schedule. Because that's how *work* works in the real world.

Dog-walking. Volunteering at a soup kitchen. Shoveling sidewalks, mowing lawns, selling greeting cards. When I was thirteen, Mother arranged for me to get a seamstress job for two afternoons a week, working for a dressmaker out of her home. I liked feeling useful. And I liked making money. This was the first time I had earned money at a job and I bought some crazy shirts with it, pullovers with stripes. (Unfortunately, Mother "lost" them in the laundry.) During high school I worked summers at my aunt's ranch. Even though I talked nonstop about topics that bored people, everyone loved the horse bridles I made.

Obsessions, in fact, can be great motivators. A creative teacher or parent can channel obsessions into career-relevant skills. If a child likes trains, read a book about trains and do math with trains. My science teacher used my obsession with my squeeze machine to motivate scientific study. He told me that if I wanted to argue that physical pressure is relaxing, I had to learn how to read scientific journal articles to support my thesis.

Not all obsessions are created equal, of course. I see kids who are so addicted to video games that you can't get them interested in anything else—though even then, I know of one parent who encouraged development of artistic ability by having her son draw pictures of video-game characters. But if you can't turn video-gaming into a learning opportunity, you can at least restrict it to one hour per day (though career-relevant skills such as programming a game can be done for much longer periods).

Just keep your eyes open for opportunities, and don't be afraid to be creative. At the grocery store the other day, I saw a magazine devoted to chickens. I started flipping through it, and I read an article about how to raise chickens in your backyard. *Now that,* I thought, *is a great opportunity for a parent.* You buy a few chickens, and suddenly a child has a "job"—or at least the opportunity to learn all sorts of skills that will be useful throughout life. You can read about chickens together, learn how to take care of them, feed them, clean up after them. The kid can even start a business—gathering the eggs, delivering them to neighbors, collecting the payments.

Of course, if you can find an opportunity that matches the child's way of thinking and that prepares the child to eventually enter the work force doing what she does best, all the better. Ideally, you want to prepare the child for employment that is not only productive but also a source of energy and joy (see sidebar at the end of this chapter).

Word-fact thinkers, for instance, would do well with writing assignments. They can contribute to the church newsletter. They can start a neighborhood blog. Maybe they can write for the local paper. After all, *somebody* has to report on how many stray dogs have been picked up that week.

Unfortunately, a lot of the jobs that are ideal for word-fact thinkers are disappearing. Filing, record-keeping, clerking—these are tasks that increasingly are being handled by computers. The trick, then, is to let the computer become the word-fact thinker's friend. A lot of these thinkers would be great at conducting elaborate Internet searches and organizing the results.

Word-fact thinkers would benefit from learning how to be what I call business-social. They can still talk, but they've got to learn when to talk and how to talk, either through getting out in the world and learning through numerous examples or through on-the-job training. Telephone sales, for instance, would be a good job for them once they've learned the script. And it's no coincidence that Leo Kanner's first patient, Donald Triplett, grew up to become a bank teller.

A *picture thinker* might be able to make art and sell it. After one of my talks recently, I met a teenage girl who designs jewelry. I know jewelry, so I can confidently say it: She has talent. She's a pro. I told her that she should sell it online, and then I told her mother how to figure out a fair price: twenty dollars per hour of labor, plus the cost of materials. At a hundred and twenty-five dollars, the bracelet I saw would be a bargain.

A *pattern thinker* who's good at math can fix computers or tutor neighborhood kids. A pattern thinker who's gifted in music can play in a band or join a choir — technically not paying jobs, usually, but jobs nonetheless, in the sense that they require cooperation with the other musicians as well as a regular commitment of time.

In short, any job that teaches autistic children about responsibilities is a job that will help prepare them for adulthood.

But job skills are only half the battle. The person with autism will also need social skills. These lessons, too, should be taught at a young age. Learning to say "Please" and "Thank you" is a basic, nuts-and-bolts necessity. So is learning to take turns; board games and card games are good instruction methods. Table manners too. Behaving appropriately in a store or a restaurant. Being on time.

Again, get those kids out into the world! The other day I talked to a mother who said that her grown-up daughter had never gone grocery shopping. Her daughter was high-functioning; she could drive a car. How will she be prepared for adult life, especially if she eventually has to live on her own, if she can't go to the store? The mother was low-income, so I told her I wasn't going to ask her to spend any money she wouldn't already be spending. "You're going to buy the groceries anyway," I said. "But have your daughter do it. Give her the shopping list, give her some money or a credit card, and send her into the store. You can wait in the parking lot."

Mother made me do social stuff I didn't want to do. I remember being scared to go to the lumberyard by myself because I was afraid to talk to the clerks. But Mother insisted. So I went, and I came back home crying. But I had the wood I wanted — plus a new social skill. Next time I could go to the lumberyard with less trepidation and greater confidence.

These basics are just the foundation — the social skills that are a given for anybody entering the work force. People with autism, however, often have to master more specialized social skills.

I remember two kids I went to school with who would be labeled as having Asperger's syndrome today. One has a PhD and a good job as a psychologist. The other has held on to good retail jobs and is a valued member of the store's staff because he can talk to customers about every product in the store. In the meat industry, I have worked with many successful individuals who are, I'm pretty sure, undiagnosed Asperger's. At one plant I visited, the undiagnosed Aspies never went in the cafeteria; instead, they ate their lunches at a picnic table in the shop. I once visited a research lab for fish farming. I could see that all the equipment was put together from materials available at Home Depot — water filters made out of window-screen mesh, for instance. The lab was amazingly inventive, so of course I had to ask whose was the mind behind all this innovation. It turned out to be the (nondiagnosed) Aspie who was the maintenance guy at the time he'd created these inventions — and who had now graduated to running the lab.

All these people were fortunate to find jobs in fields where they could flourish. Some of them, like the fish-farming lab director, had to come in the back door. But at least he knew what to do once he walked through it.

I'm not sure that would be possible today. I have talked to numerous young people with Asperger's syndrome who have been fired from their jobs. Yet their condition was no more or less severe than the kids I knew in school, or the Aspies who ate lunch together, or the fish-farming research director, or any of the other on-the-spectrum people I've met who have managed to keep their jobs for decades. It's a generational thing, I suspect. The younger generation doesn't know how to behave. Maybe the families and facilitators of kids who have received official diagnoses since the addition of ASD to the *DSM* in 1980 have become so focused on the label—and the deficits—that they think they don't need to attend to the social skills that are necessary to advance in society. I don't want to sound like some old coot who's always talking about how much better everything was way back in the good old days. But when I ask these people why they were fired, I find out that they didn't know how to do simple tasks like show up on time or that they were doing stupid things that I learned not to do when I was nine years old.

Here's my advice—the advice I give to folks who ask me how to prepare someone who's on the spectrum for employment.

- *Don't make excuses.*

A high-school senior was complaining to me that he screwed up in English class because of a learning disability, and then he mentioned that he had done well in a philosophy course. "Wait a minute," I said. "Writing an English paper and writing a philosophy paper require the same skills. Don't tell me you have a learning disability in English." He insisted he did. I kept pressing, and sure enough, he finally said he wasn't interested in English, but he did like philosophy.

First of all, "I'm not interested" isn't a good excuse for not performing a necessary task the best that you can; it just means you have to work harder than you would at the task you enjoy. But "I have a learning disability" is an even worse excuse if it's not the real reason.

- *Play well with others.*

I know one woman who was constantly getting into verbal fights—with the bus driver, the lady at the post office, you name it. Every day. And of course it was never her fault. It was always the other person who was being unreasonable. She would tell me this, and I'd think, *How do you get into a fight with a different bus driver every day? Most people don't even* talk *to the bus driver!* I hear too many individuals with Asperger's syndrome saying things like "I have authority issues with the boss." I want to tell them that there's a reason the boss is called the boss. It's because she's *the boss.*

That's a lesson I learned the hard way. I was doing a summer internship during college at a hospital that had a program for kids with autism and other problems, and my boss did something with a kid that I didn't like. I don't remember what it was, but I do remember that I went over his head. I took my complaint to the psychology department, which was a different department. My boss didn't fire me, but he did let me know he was upset. He told me about the hierarchy at the hospital, and how I worked for the child-care department, and that if I had a complaint I should go to him first. And he was right. And I never made that mistake again.

Playing well with others, however, isn't just about avoiding confrontations. It's also about learning to try to please. My mother motivated me by making sure that I got real recognition when I did a good job—like when she framed a watercolor of the beach that I'd painted. Another time, I was allowed to sing a solo at an adult concert. I was thrilled. I knew this was a special privilege, and when the audience responded with applause and cheers, I felt tremendous pride. In high school, I painted signs for many different people. I learned that when I made a sign for a hair salon, for example, I had to paint a design the client would like. These were the experiences I later drew on when I embarked on my design career. I wanted to do work that people really appreciated.

- *Manage your emotions.*

How do you that? By learning to cry. And how do you do *that?* By giving yourself permission. (And if you're in a position to give someone else that permission, then do it.) You don't have to cry in public. You don't have to cry in front of your peers. But if the alternative is to hit or throw, then, yes, you do have to cry. When parents tell me that their teenage boy cries when he's frustrated, I say, "Good!" Boys who cry can work for Google. Boys who trash computers cannot. I once was at a science conference, and I saw a NASA scientist who had just found out that his project was canceled—a project he'd worked on for years. He was maybe sixty-five years old, and you know what? He was crying. And I thought, *Good for him.* That's why he was able to reach retirement age working in a job he loved.

From a neuroscience point of view, managing emotions depends on top-down control from the frontal cortex. If you can't control your emotion, you have to *change* your emotion. If you want to keep a job, you have to learn how to turn anger into frustration. I saw in a magazine article that Steve Jobs would cry in frustration. That's why Steve Jobs still had a job. He could be verbally abusive to his employees, but as far as I know, he didn't go around throwing things at them or slugging them.

I learned my lesson in high school. I got in a fight with someone who was teasing me, and I had horseback riding taken away for two weeks. That's the last fight I ever had. When I got into the cattle business, I was angry plenty of times, but I knew enough not to show it. Instead I would hide out on the cattle catwalk. I was right in plain view, but I knew I was so far off the ground that nobody could see I was crying. Or I'd go underneath some stairs, or I'd sit in my car in the parking lot. Sometimes I'd go in the electrical room, because the lovely sign on the door told everybody else to KEEP OUT. But I'd never hide in the restroom, because I couldn't know if someone was going to walk in.

- *Mind your manners.*

When I was about eight years old, I learned that calling somebody Fatso was not appropriate. I've met a number of high-functioning autistics and Asperger's individuals who have been fired from jobs because they made rude comments about the appearance of coworkers and customers. Even if you've reached adulthood without knowing what's rude or how to relate to people in public, it's not too late to learn.

I met someone who told me that his therapist's advice on how to learn to socialize was to practice saying hello. I told him that advice isn't specific enough. I told him to divide up his grocery shopping so that he would have to go to the supermarket every day, even if he just wound up buying a can of soup. Then when he got to the cashier, he should have a simple conversation.

- *Sell your work, not yourself.*

If you can avoid the front-door interview, do so. Human resources departments are usually staffed by social people who tend to place a premium on getting along and teamwork, so they might not think a person with autism is the right fit for the workplace. They might not be able to see past the social awkwardness to an individual's hidden talents. A better strategy for getting the job might be to contact the head of the particular department you want to work in (the engineering department, the graphic design department, and so on).

People thought I was weird, but they were impressed when they saw a portfolio of my drawings and photos of completed projects. I also made sure to use attractive brochures and portfolios to sell my design services. Electronic devices today can remove a lot of the social awkwardness of showing your work and even auditioning for a job. You can attach your work as a file in an e-mail, once you establish contact with a prospective employer (but not before — no one will open an unknown sender's e-mail attachment). You can store it on a smartphone, because you never know when someone might want to see it. A verbal thinker's portfolio of writing, a picture thinker's art or crafts, a musician's recordings, even a math whiz's coding — they're all portable today.

- *Use mentors.*

When I was in high school, I was an unmotivated student who seldom studied. I saw no point in studying until Mr. Carlock instilled in me the goal of becoming a scientist. I've talked to many successful individuals with Asperger's syndrome, both diagnosed and undiagnosed, who say they became successful only because they had either a parent or a teacher instructing them — and maybe even inspiring them. For instance, young people with Asperger's or high-functioning autistics might fool around with computers, but they'll need a mentor to focus them and to help them learn programming.

Okay, let's say the autistic child has gotten an education that identified and developed his or her strengths. And let's say that child has grown up to enter a marketplace that appreciates his or her particular skill set. That's great for that person. But you know what? It's also great for society.

Not only can you have different types of thinkers doing what they do best, but you can have them doing what they do best alongside other types of thinkers who are doing what *they* do best.

When I recall collaborations in which I've participated, I can see how different kinds of thinkers worked together to create a product that was greater than the sum of its parts. I think about the work I did with a student (nonautistic) who was good at everything I was bad at. Bridget was an ace at statistics, very organized, and a wonderful data collector and record-keeper — someone I could trust to run the experiment right. One experiment we did together correlated the excitability of cattle in the squeeze chute with their weight gain. We used two observers, and they rated the cattle's behavior on a scale of

1 to 4, with 1 being calm and 4 being berserk. One day Bridget came up to me and said, "Dr. Grandin, I'm afraid we're not getting any useful results." So I reran the "film" of the experiment in my mind, and I saw that the observers seemed to have two different standards about what constituted berserk behavior. Sure enough, Bridget and I found that one of the observers had a much higher percentage of 4 ratings. I can design experiments, and I can detect flaws in the methodology, because my picture thinking allows me to see what I want the experiment to do and what the experiment has gotten wrong. But I need a pattern thinker like Bridget to run the statistical analysis and do the meticulous record-keeping of the experiment.

I think about livestock construction. The pattern thinker — the degreed engineer — does not lay out the plant. The picture thinker — the draftsman — does. Only when the draftsman has finished laying out the packaging floor and the slaughter floor and so on does the engineer get to work, calculating the roof trusses, spec'ing out the concrete, figuring out the rebar spacing. The one part of the plant that this one particular draftsman I know — me — doesn't design is the refrigeration. Why? Because it requires too much pattern thinking for me to design it properly — too much mathematics and abstract engineering. I know just enough about refrigeration to stay away from it.

And I think about Mick Jackson, the director of the HBO movie *Temple Grandin*. If you look at an earlier movie of his, the Steve Martin comedy *L.A. Story*, you'll see that it doesn't have much structure. That's because Mick is a picture thinker, not a pattern thinker. By the time he was working on my movie, he knew what his strengths were and where he needed help, so every time Mick wanted to change something in the script, he would consult with one of the writers, Christopher Monger. He was a word thinker, of course, but he was also a pattern thinker who could tell what effect each little change was going to have on the overall structure. The movie benefited enormously, I think, from being created by all three kinds of thinking.

In the previous chapter I said that once I recognized pattern thinking for what it is, I started seeing it everywhere. The same is true of examples of the way the three kinds of thinking work together. Now I can see them not only in my own experience but everywhere I look.

Reading an interview with Steve Jobs, I came across this quote: "The thing I love about Pixar is that it's exactly like the LaserWriter." *What?* The most successful animation studio in recent memory is "exactly like" a piece of technology from 1985?

He explained that when he saw the first page come out of Apple's LaserWriter — the first laser printer ever — he thought, *There's awesome amounts of technology in this box.* He knew what all the technology was, and he knew all the work that went into creating it, and he knew how innovative it was. But he also knew that the public wasn't going to care about what was inside the box. Only the product was going to matter — the beautiful fonts that he made sure were part of the Apple aesthetic. This was the lesson he applied to Pixar: You can use all sorts of new computer software to create a new kind of animation, but the public isn't going to care about anything except what's on the screen.

He was right — obviously. While he didn't use the terms *picture thinker* and *pattern thinker,* that's what he was talking about. In that moment in 1985, he realized that you needed pattern thinkers to engineer the miracles inside the box and picture thinkers to make what comes out of the box beautiful.

I haven't been able to look at an iPod or iPad or iPhone without thinking about that interview. I now understand that when Apple gets something wrong, it's because they didn't get the balance between the kinds of thinking right. The notorious antenna problem on the iPhone 4? Too much art, not enough engineering.

Contrast this philosophy with Google's; the minds behind Google, I guarantee you, were pattern thinkers. And to this day, Google products favor engineering over art.

What all these examples tell me is that in society, the three kinds of minds naturally complement one another. Society puts them together without anybody thinking about it. But what if we did think about it? What if we recognized these categories consciously and tried to make the various pairings work to our advantage? What if each of us was able to say, *Oh, here's my strength, and here's my weakness — what can I do for you, and what can you do for me?*

When Richard and I started collaborating on this book, we both recognized that we worked well together. But as we developed the idea of brains being wired for different ways of thinking, we realized *why* we worked well together. Richard's a pattern and word thinker, and I'm a picture thinker. And because we realized how we complement each other's strengths, we have been able to exploit them to a greater extent than would have otherwise been possible.

I'm always saying to Richard, "You're the structure guy" — meaning that his strength in organizing the concepts in the book compensates for my weakness in that area. When I look back at papers I wrote in the 1990s, I'm embarrassed at how randomly organized they were. Concepts didn't follow concepts in logical formation. They just sort of clumped here and there — pretty much wherever they occurred to me in the process of writing the paper. I've gotten better at structure over the years, but I know I'll never be like Richard. When he tells me that a particular concept we've been chewing over belongs in chapter 6, I say, "Okay."

Fine. Good for us. Even if I weren't autistic, we'd be a good team, because our kinds of minds complement each other. But the fact is, I am autistic, and the strengths I bring to the collaboration are strengths that belong to my kind of autistic brain — the quick associations, the long-term memory, the focus on details.

Let's apply this same principle to the marketplace. If people can consciously recognize the strengths and weaknesses in their ways of thinking, they can then seek out the right kinds of minds for the right reasons. And if they do that, then they're going to recognize that sometimes the right mind can belong only to an autistic brain.

We've discussed how autistic brains seem to be better at picking out details than normal brains. If we see that kind of trait not as a byproduct of bad wiring but simply as the product of wiring—the kind of argument that Michelle Dawson made in chapter 6—then we can begin to see it as offering a possible advantage in some circumstances. And if we see that being able to see the trees before the forest might make someone better at seeing certain kinds of patterns, then we can ask where that skill might be useful. And if we realize that security screeners at the airport need to pick out details quickly, then there we go: a job.

By cultivating the autistic mind on a brain-by-brain, strength-by-strength basis, we can reconceive autistic teens and adults in jobs and internships not as charity cases but as valuable, even essential, contributors to society.

Some entrepreneurs have already made that leap. Aspiritech, in the Chicago suburb of Highland Park, and Specialisterne, in Copenhagen, both employ primarily high-functioning autistics and individuals with Asperger's to test software. Their brains—wired to endure repetition, to focus closely, to remember details—are just what the job requires. The son of Aspiritech's founder was diagnosed with Asperger's at the age of fourteen, and as an adult he was fired from his job as a grocery bagger. But when it comes to testing software, he's the go-to guy.

In 2007 Walgreens opened a distribution center in Anderson, South Carolina, that hired a work force that was 40 percent persons with disabilities, including those on the autism spectrum. The idea was the brainchild of Randy Lewis, a vice president at the retailer who was the father of an autistic son. Thanks to touchscreens and flexible workstations, the employees with disabilities work side by side with their "normal" peers. When Walgreens saw that the center was 20 percent more efficient than the company's other centers, it expanded the philosophy to another distribution center, in Windsor, Connecticut, in 2009.

But you don't need to wait for a big corporation with an enlightened hiring policy to build a branch near you. Parents can take their autistic kid to a neighborhood shop or restaurant, talk to the owner or manager, and see if there might be a job available that would be suitable for the child's skill level. And if one door closes, and another, and even another, "keep on knockin'."

That advice is courtesy of Savino Nuccio D'Argento—Nuccio to everyone. He (along with a business partner) owns Vince's Italian Restaurant in the Chicago suburb of Harwood Heights. Nuccio has an autistic son, Enzo, and through his contacts with the autism branch of Chicago Easter Seals, Nuccio regularly hires adult persons with autism. He also opens his doors to training programs for school-age kids; they learn to vacuum, set tables, make sure the salt and pepper shakers are filled—the kinds of tasks that will help prepare some of them for entry into the adult world.

"For other people, it would be like, 'Oh, I hate this job,'" Nuccio says. Not so for people with autism. "They love it, because every day it's the same thing."

The problems he's encountered, in fact, haven't come from the autistic employees and trainees. Instead, he says, they've come from the "normal" employees who resist the change in their work environment.

"It still takes time for other people to accept it," he says. "There are still people out there who look at it as, 'Oh, heck, I've got to deal with this.' It's sad. It saddened me at first because I didn't think I had employees who thought that way. But you've just got to get them to cross that hurdle and let them know it'll be okay." Maybe the first couple of weeks are rough on the other employees, he says, and he understands why. "They've got to deal with this person asking them the same question over and over and over again." In the end, though, the employees adjust—especially, he says, once they have an epiphany: "We're helping these people, sure, but they're going to end up helping us, because they're going to do their job really well."

If necessary, Easter Seals will try to place the trainees in paying positions elsewhere. One trainee went on to answer phones for Easter Seals. Another works forty-hour weeks at a produce store. Nuccio hope his own son, now fourteen, will one day reach the same happy outcome — happy for both of them. As Randy Lewis, the Walgreens executive, told NBC News, the inspiration for his hiring innovation was the age-old question that haunts so many parents of children with disabilities: *What will happen to my child when I'm gone?* To which the mother of an adult with Asperger's who worked at the distribution center in Anderson answered: "I don't have that worry anymore."

And what about the employees themselves — the people with autism who are fortunate enough to knock on the right door? Here's an inspirational case that recently came to my attention.

In the fall of 2009, John Fienberg, a high-functioning autistic, got a temp job at a New York City ad agency as a digital librarian — a great gig for a word thinker like John. It was supposed to last only a week, but John's skills — accuracy, speed, and a willingness to perform repetitive tasks that vexed normal brains — made him a valuable asset to the agency. He continued temping there for six months, until the company found money in the budget to hire him full-time. Today he catalogues, files, and otherwise manages the product photography, advertising masters, and stock imagery in the ad firm's digital library.

"I am naturally very detail-oriented in a way that makes cataloguing very easy for me," he wrote in an e-mail. The fact that he was communicating via e-mail was a reflection of his social skills. When we contacted him by e-mail (Richard heard about him through a friend), he said that he would be willing to be interviewed, but that he strongly preferred not to talk over the phone. He also said that meeting in person could be a problem; he knows he exhausts people with his over-talking.

"My boss is aware of my disabilities and does his best to work with me," John continued, "and I try to repay him by producing results that make it worth putting up with me when I don't quite understand something the way he wishes I would. The rest of my coworkers do not interact with me except for the phone and through e-mails." Still, he said, "to the best of my knowledge they all really like me and appreciate my contributions. I even got a commendation from one of them last month that was shared at the staff meeting."

John is 29 now and recently engaged. He and his fiancée plan to leave New York for "somewhere where the money I get goes further." Don't worry, though, about whether he can find another job that's such a great match. "I have permission from work to telecommute permanently."

We've come a long way from the days of doctors telling the parents of autistic children that the situation was hopeless and that the only humane option was a life sentence in an institution.

We have a lot farther to go, of course. Ignorance and misunderstanding are always difficult to overcome when they've become part of a society's belief system. For instance, when the movie *The Social Network* came out, in 2010, the *New York Times* op-ed columnist David Brooks wrote this assessment of the onscreen character of Mark Zuckerberg, the founder of Facebook: "It's not that he's a bad person. He's just never been house-trained." The "training" of the fictional character, however, would have had to somehow accommodate a brain that can't process facial and gestural cues that most people easily assimilate and that finds its greatest fulfillment not in the fizzy buzz of forming a personal relationship but in the click-clack logic of writing code.

When something is "all in your mind," people tend to think that it's willful, that it's something you could control if only you tried harder or if you had been trained differently. I'm hoping that the newfound certainty that autism is in your brain and in your genes will affect public attitudes.

As we've seen, it's already affecting research, prompting scientists to redouble efforts to look for cause and cure. And it's already affecting therapeutic attitudes, shifting the emphasis from a sole focus on deficits to a broader appreciation of strengths.

When I look back on where autism was sixty years ago, when my autistic brain was creating great anxiety in Mother, curiosity in doctors, and a challenge to my nanny and teachers, I know that trying to imagine where we'll be sixty years from now is a fool's errand. But I have confidence that whatever the thinking about autism is, it will incorporate a need to consider it brain by brain, DNA strand by DNA strand, trait by trait, strength by strength, and, maybe most important of all, individual by individual.

Jobs for Picture Thinkers

- Architectural and engineering drafter
- Photographer
- Animal trainer
- Graphic artist
- Jewelry/crafts designer
- Web designer
- Veterinary technician
- Auto mechanic
- Machine maintenance technician
- Computer troubleshooter
- Theater lighting director
- Industrial automation designer
- Landscape designer
- Biology teacher
- Satellite map analyst
- Plumber
- Heating, ventilation, and air-conditioning technician

- Photocopier repair technician
- Audio/visual equipment technician
- Welder
- Plant engineer
- Radiological technician
- Medical-equipment repair technician
- Industrial designer
- Computer animator

Jobs for Word-Fact Thinkers

- Journalist
- Translator
- Specialty retailer (that is, a worker in a store that sells only one kind of product)
- Librarian
- Stocks and bonds analyst
- Copyeditor
- Accountant
- Budget analyst
- Bookkeeper and record-keeper
- Special-education teacher
- Book indexer
- Speech therapist
- Inventory-control specialist
- Legal researcher
- Contract specialist for auto dealership
- Historian
- Technical writer
- Bank teller
- Tour guide
- Person at an information counter

Jobs for Pattern Thinkers

- Computer programmer
- Engineer

- Physicist
- Musician/composer
- Statistician
- Math teacher
- Chemist
- Electronics technician
- Music teacher
- Scientific researcher
- Mathematical data mining analyst
- Stock and financial investing analyst
- Actuary
- Electrician

Appendix: The AQ Test

PSYCHOLOGIST SIMON BARON-COHEN and his colleagues at Cambridge's Autism Research Centre have created the Autism-Spectrum Quotient, or AQ, as a measure of the extent of autistic traits in adults. In the first major trial using the test, the average score in the control group was 16.4. Eighty percent of those diagnosed with autism or a related disorder scored 32 or higher. The test is not a means for making a diagnosis, however, and many who score above 32 and even meet the diagnostic criteria for mild autism or Asperger's report no difficulty functioning in their everyday lives.

	Definitely agree	**Slightly agree**	**Slightly disagree**	**Definitely disagree**
1. I prefer to do things with others rather than on my own.				
2. I prefer to do things the same way over and over again.				
3. If I try to imagine something, I find it very easy to create a picture in my mind.				

	Definitely agree	Slightly agree	Slightly disagree	Definitely disagree
4. I frequently get so strongly absorbed in one thing that I lose sight of other things.				
5. I often notice small sounds when others do not.				
6. I usually notice car number plates or similar strings of information.				
7. Other people frequently tell me that what I've said is impolite, even though I think it is polite.				
8. When I'm reading a story, I can easily imagine what the characters might look like.				

	Definitely agree	Slightly agree	Slightly disagree	Definitely disagree
9. I am fascinated by dates.				
10 In a social group, I can easily keep track of several different people's conversations.				
11. I find social situations easy.				
12. I tend to notice details that others do not.				
13. I would rather go to a library than to a party.				
14. I find making up stories easy.				

	Definitely agree	Slightly agree	Slightly disagree	Definitely disagree
15. I find myself drawn more strongly to people than to things.				
16. I tend to have very strong interests, which I get upset about if I can't pursue.				
17. I enjoy social chitchat.				
18. When I talk, it isn't always easy for others to get a word in edgewise.				
19. I am fascinated by numbers.				
20. When I'm reading a story, I find it difficult to work out the characters' intentions.				

	Definitely agree	Slightly agree	Slightly disagree	Definitely disagree
21. I don't particularly enjoy reading fiction.				
22. I find it hard to make new friends.				
23. I notice patterns in things all the time.				
24. I would rather go to the theater than to a museum.				
25. It does not upset me if my daily routine is disturbed.				
26. I frequently find that I don't know how to keep a conversation going.				

	Definitely agree	Slightly agree	Slightly disagree	Definitely disagree
27. I find it easy to "read between the lines" when someone is talking to me.				
28. I usually concentrate more on the whole picture, rather than on the small details.				
29. I am not very good at remembering phone numbers.				
30. I don't usually notice small changes in a situation or a person's appearance.				
31. I know how to tell if someone listening to me is getting bored.				

	Definitely agree	Slightly agree	Slightly disagree	Definitely disagree
32. I find it easy to do more than one thing at once.				
33. When I talk on the phone, I'm not sure when it's my turn to speak.				
34. I enjoy doing things spontaneously.				
35. I am often the last to understand the point of a joke.				
36. I find it easy to work out what someone is thinking or feeling just by looking at their face.				
37. If there is an interruption, I can switch back to what I was doing very quickly.				

	Definitely agree	Slightly agree	Slightly disagree	Definitely disagree
38. I am good at social chitchat.				
39. People often tell me that I keep going on and on about the same thing.				
40. When I was young, I used to enjoy playing games involving pretending with other children.				
41. I like to collect information about categories of things (e.g., types of cars, birds, trains, plants).				
42. I find it difficult to imagine what it would be like to be someone else.				

	Definitely agree	Slightly agree	Slightly disagree	Definitely disagree
43. I like to carefully plan any activities I participate in.				
44. I enjoy social occasions.				
45. I find it difficult to work out people's intentions.				
46. New situations make me anxious.				
47. I enjoy meeting new people.				
48. I am a good diplomat.				

	Definitely agree	Slightly agree	Slightly disagree	Definitely disagree
49. I am not very good at remembering people's date of birth.				
50. I find it very easy to play games with children that involve pretending.				

How to score: "Definitely agree" or "Slightly agree" responses to questions 2, 4, 5, 6, 7, 9, 12, 13, 16, 18, 19, 20, 21, 22, 23, 26, 33, 35, 39, 41, 42, 43, 45, and 46 score 1 point. "Definitely disagree" or "Slightly disagree" responses to questions 1, 3, 8, 10, 11, 14, 15, 17, 24, 25, 27, 28, 29, 30, 31, 32, 34, 36, 37, 38, 40, 44, 47, 48, 49, and 50 score 1 point.

Notes

1. The Meanings of Autism

5 A few years earlier: John Donvan and Caren Zucker, "Autism's First Child," *Atlantic,* October 2010.

6 that 1943 paper: Leo Kanner, "Autistic Disturbances of Affective Contact," *Nervous Child* 2 (1943): 217–50.

a 1949 follow-up: Leo Kanner, "Problems of Nosology and Psychodynamics in Early Childhood Autism," *American Journal of Orthopsychiatry* 19, no. 3 (1949): 416–26.

7 interview in *Time*: "Medicine: The Child Is Father," *Time,* July 25, 1960, http://autismedsp5310s20f10.pbworks.com/f/Time-The+Child+Is+Father.pdf.

"misquoted often": http://www.autism-help.org/points-refrigerator-mothers.htm.

8 "If Temple doesn't": Eustacia Cutler, *Thorn in My Pocket: Temple Grandin's Mother Tells the Family Story* (Arlington, TX: Future Horizons, 2004).

footnote: Richard Pollak, *The Creation of Dr. B: A Biography of Bruno Bettelheim* (New York: Simon & Schuster, 1997).

10 several standard tests: Temple Grandin, "My Experiences as an Autistic Child and Review of Selected Literature," *Journal of Orthomolecular Psychiatry* 13, no. 3 (1984): 144–74.

13 This reversal happened: Roy Richard Grinker, *Unstrange Minds: Remapping the World of Autism* (New York: Basic Books, 2007).

published a paper: D. L. Rosenhan, "On Being Sane in Insane Places," *Science* 179, no. 4070 (January 19, 1973): 250–58.

14 A 1996 study: Lynn Waterhouse et al., "Diagnosis and Classification in Autism," *Journal of Autism and Developmental Disorders* 26, no. 1 (1996): 59–86.

15 Lorna Wing: Lorna Wing, "Asperger's Syndrome: A Clinical Account," *Psychological Medicine* 11 (1981): 115–30.

16 Columbia University study: Marissa King and Peter Bearman, "Diagnostic Change and the Increased Prevalence of Autism," *International Journal of Epidemiology* 38, no. 5 (October 2009): 1224–34.

17 Columbia University analysis: Ka-Yuet Liu, Marissa King, and Peter Bearman, "Social Influence and the Autism Epidemic," *American Journal of Sociology* 115, no. 5 (March 2010): 1387–1434.

18 another reason: Grinker, *Unstrange Minds*.

19 (ADDM) Network: http://www.cdc.gov/ncbddd/autism/addm.html.

"There's a long": Jeffrey S. Anderson interview.

2. Lighting Up the Autistic Brain

27 Thanks to a scan: Eric Courchesne et al., "Cerebellar Hypoplasia and Hyperplasia in Infantile Autism," *Lancet* 343, no. 8888 (January 1, 1994): 63–64.

28 I found one: N. Shinoura et al., "Impairment of Inferior Longitudinal Fasciculus Plays a Role in Visual Memory Disturbance," *Neurocase* 13, no. 2 (April 2007): 127–30.

32 "The entorhinal cortex": http://newsroom.ucla.edu/portal/ucla/ucla-scientists-boost-memory-by-228557.aspx.

33 The latest estimate: Sarah DeWeerdt, "Study Links Brain Size to Regressive Autism," Simons Foundation Autism Research Initiative, December 12, 2011, http://sfari.org/news-and-opinion/news/2011/study-links-brain-size-to-regressive-autism.

34 one review article: Nancy J. Minshew and Timothy A. Keller, "The Nature of Brain Dysfunction in Autism: Functional Brain Imaging Studies," *Current Opinion in Neurology* 23, no. 2 (April 2010): 124–30.

"Anatomically, these kids": Joy Hirsch interview.

told USA Today: Liz Szabo, "Autism Science Is Moving 'Stunningly Fast,'" USA Today, April 30, 2012, http://usatoday30.usatoday.com/news/health/story/2012-04-08/Autism-science-research-moving-faster/54134028/1.

35 2011 fMRI study: Naomi B. Pitskel et al., "Brain Mechanisms for Processing Direct and Averted Gaze in Individuals with Autism," *Journal of Autism and Developmental Disorders* 41, no. 12 (December 2011): 1686–93.

36 highly influential paper: Marcel Adam Just et al., "Cortical Activation and Synchronization During Sentence Comprehension in High-Functioning Autism: Evidence of Underconnectivity," *Brain* 127, no. 8 (August 2004): 1811–21.

a 2012 study: M. E. Vissers et al., "Brain Connectivity and High Functioning Autism: A Promising Path of Research That Needs Refined Models, Methodological Convergence, and Stronger Behavioral Links," *Neuroscience and Biobehavioral Reviews* 36, no. 1 (January 2012): 604–25.

37 a 2009 autism study: H. C. Hazlett et al., "Teasing Apart the Heterogeneity of Autism: Same Behavior, Different Brains in Toddlers with Fragile X Syndrome and Autism," *Journal of Neurodevelopmental Disorders* 1, no. 1 (March 2009): 81–90.

38 a study her group conducted: Grace Lai et al., "Speech Stimulation During Functional MR Imaging as a Potential Indicator of Autism," *Radiology* 260, no. 2 (August 2011): 521–30.

39 a major study: Jeffrey S. Anderson et al., "Functional Connectivity Magnetic Resonance Imaging Classification of Autism," *Brain* 134 (December 2011): 3742–54.

40 A 2011 MRI study: A. Elnakib et al., "Autism Diagnostics by Centerline-Based Shape Analysis of the Corpus Callosum," *IEEE International Symposium on Biomedical Imaging: From Nano to Macro* (March 30, 2011): 1843–46.

another MRI study from 2011: Lucina Q. Uddin et al., "Multivariate Searchlight Classification of Structural Magnetic Resonance Imaging in Children and Adolescents with Autism," *Biological Psychiatry* 70, no. 9 (November 2011): 833–41.

41 a 2012 DTI study: Jason J. Wolff et al., "Differences in White Matter Fiber Tract Development Present from 6 to 24 Months in Infants with Autism," *American Journal of Psychiatry* 169, no. 6 (June 2012): 589–600.

42 "They came to me": Walter Schneider interview.

posted a paper: S. S. Shin et al., "High-Definition Fiber Tracking for Assessment of Neurological Deficit in a Case of Traumatic Brain Injury: Finding, Visualizing, and Interpreting Small Sites of Damage," *Journal of Neurosurgery* 116, no. 5 (May 2012): 1062–69.

45 my book *Emergence*: Temple Grandin and Margaret M. Scariano, *Emergence* (New York: Warner Books, 1996).

47 "It really, really": Virginia Hughes, "Movement During Brain Scans May Lead to Spurious Patterns," Simons Foundation Autism Research Initiative, January 16, 2012, http://sfari.org/news-and-opinion/news/2012/movement-during-brain-scans-may-lead-to-spurious-patterns.

48 an article in *Science:* Greg Miller, "Growing Pains for fMRI," *Science* 320 (June 13, 2008): 1412–14.

3. Sequencing the Autistic Brain

50 "The human genome": Gina Kolata, "Study Discovers Road Map of DNA," *New York Times*, September 6, 2012.

52 The article: Amartya Sanyal et al., "The Long-Range Interaction Landscape of Gene Promoters," *Nature* 489 (September 6, 2012): 109–13.

53 the first study of autism in twins: S. Folstein and M. Rutter, "Infantile Autism: A Genetic Study of 21 Twin Pairs," *Journal of Child Psychology and Psychiatry* 18, no. 4 (September 1977): 297–321.

54 A follow-up study: A. Bailey et al., "Autism as a Strongly Genetic Disorder: Evidence from a British Twin Study," *Psychological Medicine* 25, no. 1 (January 1995): 63–77.

55 Autism Genome Project, or AGP: http://www.autismspeaks.org/science/initiatives/autism-genome-project/first-findings.

came to an end: http://www.autismspeaks.org/about-us/press-releases/autism-speaks-and-worlds-leading-autism-experts-announce-publication-autism-.

a paper in *Nature Genetics:* Peter Szatmari et al., "Mapping Autism Risk Loci Using Genetic Linkage and Chromosomal Rearrangements," *Nature Genetics* 39, no. 3 (March 2007): 319–28.

56 a 2007 study: Jonathan Sebat et al., "Strong Association of De Novo Copy Number Mutations with Autism," *Science* 316, no. 5823 (April 20, 2007): 445–49.

an end, in 2010: http://www.autismspeaks.org/about-us/press-releases/new-autism-genes-discovered-autism-speaks-and-worlds-leading-autism-experts.

57 "We found many": http://geschwindlab.neurology.ucla.edu/index.php/in-the-news/16-news/88-dna-scan-for-familial-autism-finds-variants-that-disrupt-gene-activity-in-autistic-kids-.

an article in *Science:* Matthew W. State and Nenad Šestan, "The Emerging Biology of Autism Spectrum Disorders," *Science* 337 (September 2012): 1301–3.

58 "The key is trying": G. Bradley Schaefer interview.

One of those studies: Stephen Sanders et al., "De Novo Mutations Revealed by Whole-Exome Sequencing Are Strongly Associated with Autism," *Nature* 485 (May 10, 2012): 237–41.

At the same time, another study: Brian J. O'Roak et al., "Sporadic Autism Exomes Reveal a Highly Interconnected Protein Network of De Novo Mutations," *Nature* 485 (May 10, 2012): 246–50.

Then a third study: Benjamin M. Neale et al., "Patterns and Rates of Exonic De Novo Mutations in Autism Spectrum Disorders," *Nature* 485 (May 10, 2012): 242–45.

59 a paper in *Nature:* Augustine Kong et al., "Rate of De Novo Mutations and the Importance of Father's Age to Disease Risk," *Nature* 488 (August 2012): 471–75.

"The development of the brain": Deborah Rudacille, "Family Sequencing Study Boosts Two-Hit Model of Autism," Simons Foundation Autism Research Initiative, May 15, 2011, http://sfari.org/news-and-opinion/news/2011/family-sequencing-study-boosts-two-hit-model-of-autism.

a 2012 analysis: Claire S. Leblond et al., "Genetic and Functional Analyses of SHANK2 Mutations Suggest a Multiple Hit Model of Autism Spectrum Disorders," *PLoS Genetics* 8, no. 2 (February 2012): e1002521, doi:10.1371/journal.pgen.1002521.

60 "For these patients": Virginia Hughes, "SHANK2 Study Bolsters 'Multi-Hit' Gene Model of Autism," Simons Foundation Autism Research Initiative, February 13, 2012, http://sfari.org/news-and-opinion/news/2012/shank2-study-bolsters-multi-hit-gene-model-of-autism.

61 "It is widely accepted": http://www.universityofcalifornia.edu/news/article/25624.

"We expect to find": http://www.universityofcalifornia.edu/news/article/24693.

Hertz-Picciotto says: Irva Hertz-Picciotto interview.

The first CHARGE study: R. J. Schmidt et al., "Prenatal Vitamins, One-Carbon

Metabolism Gene Variants, and Risk for Autism," *Epidemiology* 22, no. 4 (July 2011): 476–85.

Another CHARGE study: H. E. Volk et al., "Residential Proximity to Freeways and Autism in the CHARGE Study," *Environmental Health Perspectives* 119, no. 6 (June 2011): 873–77.

A third CHARGE study: P. Krakowiak et al., "Maternal Metabolic Conditions and Risk for Autism and Other Neurodevelopmental Disorders," *Pediatrics* 129, no. 5 (May 2012): 1121–28.

62 another 2012 paper: J. F. Shelton et al., "Tipping the Balance of Autism Risk: Potential Mechanisms Linking Pesticides and Autism," *Environmental Health Perspectives* 120, no. 7 (April 2012): 944–51.

an editorial: Philip J. Landrigan et al., "A Research Strategy to Discover the Environmental Causes of Autism and Neurodevelopmental Disabilities," *Environmental Health Perspectives* 120, no. 7 (July 2012): a258–a260.

a safety alert: http://www.fda.gov/Safety/MedWatch/SafetyInformation/SafetyAlertsforHumanMedicalProducts/ucm261610.htm.

63 two studies: Miriam E. Tucker, "Valproate Exposure Associated with Autism, Lower IQ," Internal Medicine News Digital Network, December 5, 2011, http://www.internalmedicinenews.com/specialty-focus/women-s-health/single-article-page/valproate-exposure-associated-with-autism-lower-iq.

"An estimated six": Simons Foundation Autism Research Initiative, June 5, 2012, https://sfari.org/news-and-opinion/blog/2012/valproate-fate.

The first study: Lisa A. Croen et al., "Antidepressant Use During Pregnancy and Childhood Autism Spectrum Disorders," *Archives of General Psychiatry* 68, no. 11 (November 2011): 1104–12.

64 Along comes a study: A. J. Wakefield et al., "Ileal-Lymphoid-Nodular Hyperplasia, Non-Specific Colitis, and Pervasive Developmental Disorder in Children," *Lancet* 351, no. 9103 (February 28, 1998): 637–41.

The Lancet retracts: Editors of *The Lancet*, "Retraction — 'Ileal-Lymphoid-Nodular Hyperplasia, Non-Specific Colitis, and Pervasive Developmental Disorder in Children,'" *Lancet* 375, no. 9713 (February 6, 2010): 445.

65 more compelling example: David Dobbs, "The Orchid Children," *New Scientist*, January 28, 2012.

66 a neurotransmitter: http://www.utexas.edu/research/asrec/dopamine.html.

a study published in 2010: Kenneth D. Gadow et al., "Parent-Child DRD4 Genotype as a Potential Biomarker for Oppositional, Anxiety, and Repetitive Behaviors in Children with Autism Spectrum Disorder," *Progress in Neuro-Psychopharmacology and Biological Psychiatry* 34, no. 7 (October 1, 2010): 1208–14.

67 think to ask: J. Belsky et al., "Vulnerability Genes or Plasticity Genes?" *Molecular Psychiatry* 14, no. 8 (August 2009): 746–54.

"orchid children": W. Thomas Boyce and Bruce J. Ellis, "Biological Sensitivity to

Context: I. An Evolutionary–Developmental Theory of the Origins and Functions of Stress Reactivity," *Development and Psychopathology* 17, no. 2 (June 1, 2005): 271–301.

68 "We must recollect": Sigmund Freud, "On Narcissism: An Introduction," in *The Standard Edition of the Complete Psychological Works of Sigmund Freud,* vol. 14 (London: Hogarth Press, 1957).

"The deficiencies": Sigmund Freud, "Beyond the Pleasure Principle," in *The Standard Edition of the Complete Psychological Works of Sigmund Freud,* vol. 18 (London: Hogarth Press, 1955).

4. Hiding and Seeking

71 a 2011 review article: Elysa Jill Marco et al., "Sensory Processing in Autism: A Review of Neurophysiologic Findings," *Pediatric Research* 69, no. 5, pt. 2 (May 2011): 48R–54R.

72 one 2009 study: Laura Crane et al., "Sensory Processing in Adults with Autism Spectrum Disorders," *Autism* 13, no. 3 (May 2009): 215–28.

another study that same year: Lisa D. Wiggins et al., "Brief Report: Sensory Abnormalities as Distinguishing Symptoms in Autism Spectrum Disorders in Young Children," *Journal of Autism and Developmental Disorders* 39 (2009): 1087–91.

big scholarly book: David Amaral et al., eds., *Autism Spectrum Disorders* (New York: Oxford University Press, 2011).

"like someone is drilling": http://www.autismsouthafrica.org/virtual library.htm. "Asperger adults describe their experience of sensory overload."

one study indicates: B. A. Corbett et al., "Cortisol Circadian Rhythms and Response to Stress in Children with Autism," *Psychoneuroendocrinology* 31, no. 1 (January 2006): 59–68.

74 published a paper: A. E. Lane et al., "Sensory Processing Subtypes in Autism: Association with Adaptive Behavior," *Journal of Autism Developmental Disorders* 40, no. 1 (January 2010): 112–22.

76 article in *Physical Therapy:* Anjana N. Bhat, "Current Perspectives on Motor Functioning in Infants, Children, and Adults with Autism Spectrum Disorders," *Physical Therapy* 91, no. 7 (July 2011): 1116–29.

78 his book: Tito Rajarshi Mukhopadhyay, *How Can I Talk If My Lips Don't Move: Inside My Autistic Mind* (New York: Arcade Publishing, 2008).

80 her 2012 book: Arthur Fleischmann and Carly Fleischmann, *Carly's Voice: Breaking Through Autism* (New York: Touchstone, 2012).

84 One paper, published: Henry Markram, "The Intense World Syndrome—an Alternative Hypothesis for Autism," *Frontiers in Neuroscience* 1, no. 1 (2007): 77–96.

Another paper: B. Gepner and F. Féron, "Autism: A World Changing Too Fast for a Mis-Wired Brain?," *Neuroscience and Biobehavioral Reviews* 33, no. 8 (September 2009): 1227–42.

87 Which is not to say: Temple Grandin, "Visual Abilities and Sensory Differences in a Person with Autism," *Biological Psychiatry* 65 (2009): 15–16.

88 "Light refraction": Donna Williams, *Autism: An Inside-Out Approach* (London: Jessica Kingsley Publishers, 1996).

"Picasso vision": http://www.autismathomeseries.com/library/2009/08/inside-the-mind-of-sensory-overload/.

90 "I can't tolerate": http://www.wrongplanet.net/postp4758182.html&highlight=.

"You may have to": http://thewildeman2.hubpages.com/hub/Autistic-Sensory-Overload.

91 A 2003 study: Nathalie Boddaert et al., "Perception of Complex Sounds: Abnormal Pattern of Cortical Activation in Autism," *American Journal of Psychiatry* 160, no. 11 (2003): 2057–60.

Another study from 2003: F. Tecchio et al., "Auditory Sensory Processing in Autism: A Magnetoencephalographic Study," *Biological Psychiatry* 54, no. 6 (September 2003): 647–54.

a 2012 fMRI study: Sandra Sanchez, "Functional Connectivity of Sensory Systems in Autism Spectrum Disorders: An fcMRI study of Audio-Visual Processing" (PhD diss., San Diego State University, 2011).

have long noted: See, for example, I. Molnar-Szakacs and P. Heaton, "Music: A Unique Window into the World of Autism," *Annals of the New York Academy of Sciences* 1252 (April 2012): 318–24.

In a 2012 study: Grace Lai et al., "Neural Systems for Speech and Song in Autism," *Brain* 135, no. 3 (March 2012): 961–75.

92 A 2005 study: R. S. Kaplan and A. L. Steele, "An Analysis of Music Therapy Program Goals and Outcomes for Clients with Diagnoses on the Autism Spectrum," *Journal of Music Therapy* 42, no. 1 (Spring 2005): 2–19.

a 2010 paper: Catherine Y. Wan and Gottfried Schlaug, "Neural Pathways for Language in Autism: The Potential for Music-Based Treatments," *Future Neurology* 5, no. 6 (2010): 797–805.

93 a proof-of-concept study: Catherine Y. Wan et al., "Auditory-Motor Mapping Training as an Intervention to Facilitate Speech Output in Non-Verbal Children with Autism: A Proof of Concept Study," *PLoS ONE* 6, no. 9 (2011): e25505, doi:10.1371/journal.pone.0025505.

5. Looking Past the Labels

104 A 2011 article: Lizzie Buchen, "Scientists and Autism: When Geeks Meet," *Nature* 479 (November 2011): 25–27.

autism-spectrum quotient questionnaire: Simon Baron-Cohen et al., "The Autism-Spectrum Quotient (AQ): Evidence from Asperger Syndrome/High-Functioning Autism, Males and Females, Scientists and Mathematicians," *Journal of Autism and Developmental Disorders* 31 (2001): 5–17.

106 common in people with autism: T. Buie et al., "Evaluation, Diagnosis, and Treatment of Gastrointestinal Disorders in Individuals with ASDs: A Consensus Report, " *Pediatrics* 125, supplement 1 (January 2010): S1–18.

"One of the curses": David R. Simmons et al., "Vision in Autism Spectrum Disorders," *Vision Research* 49 (2009): 2705–39.

107 In a 2010 presentation: http://iacc.hhs.gov/events/2010/slides_susan_swedo_043010.pdf.

108 researchers have shown: See, for example, K. K. Chadman, "Fluoxetine but Not Risperidone Increases Sociability in the BTBR Mouse Model of Autism," *Pharmacology, Biochemistry, and Behavior* 97, no. 3 (January 2011): 586–94.

109 A 2011 paper: Laura Pina-Camacho et al., "Autism Spectrum Disorder: Does Neuroimaging Support the *DSM-5* Proposal for a Symptom Dyad? A Systematic Review of Functional Magnetic Resonance Imaging and Diffusion Tensor Imaging Studies," *Journal of Autism and Developmental Disorders* 42, no. 7 (July 2012): 1326–41.

110 intermittent explosive disorder: See, for example, Emil F. Coccaro, "Intermittent Explosive Disorder as a Disorder of Impulsive Aggression for *DSM-5*," *American Journal of Psychiatry* 169 (June 2012): 577–88.

112 A 2012 survey: James C. McPartland et al., "Sensitivity and Specificity of Proposed *DSM-5* Diagnostic Criteria for Autism Spectrum Disorder," *Journal of the American Academy of Child and Adolescent Psychiatry* 51, no. 4 (April 2012): 368–83.

A later study: M. Huerta et al., "Application of *DSM-5* Criteria for Autism Spectrum Disorder to Three Samples of Children with *DSM-IV* Diagnoses of Pervasive Developmental Disorders," *American Journal of Psychiatry* 10 (October 2012): 1056–64.

115 A 2010 article: Judith S. Verhoeven et al., "Neuroimaging of Autism," *Neuroradiology* 52, no. 1 (2010): 3–14.

116 In a 2012 article: Matthew W. State and Nenad Šestan, "The Emerging Biology of Autism Spectrum Disorders," *Science* 337 (September 2012): 1301–3.

6. Knowing Your Own Strengths

117 According to Laurent Mottron: Laurent Mottron, "Changing Perceptions: The Power of Autism," *Nature* 479 (November 2011): 33–35.

118 A 2009 report: Grant K. Plaisted and G. Davis, "Perception and Apperception in Autism: Rejecting the Inverse Assumption," *Philosophical Transactions of the Royal Society B: Biological Sciences* 364, no. 1522 (May 2009): 1393–98.

designed a study: M. Dawson et al., "The Level and Nature of Autistic Intelligence," *Psychological Science* 18, no. 8 (August 2007): 647–62.

119 "Scientists working in autism": David Wolman, "The Autie Advantage," *New Scientist* 206 (April 2010): 32–35.

"Would you like": Madhusree Mukerjee, "A Transparent Enigma," *Scientific American*, June 2004.

120 "When a person with autism": Virginia Hughes, "Autism Often Accompanied by 'Super Vision,' Studies Find," Simons Foundation Autism Research Initiative, February 12, 2009, http://sfari.org/news-and-opinion/news/2009/autism-often-accompanied-by-super-vision-studies-find.

A landmark study: Tim Langdell, "Recognition of Faces: An Approach to the Study of Autism," *Journal of Child Psychology and Psychiatry and Allied Disciplines* 19, no. 3 (July 1978): 255–68.

121 Studies have repeatedly: See, for example, P. Murphy et al., "Perception of Biological Motion in Individuals with Autism Spectrum Disorder," *Perception 37 ECVP Abstract Supplement* (2008): 113; Evelien Nackaerts, "Recognizing Biological Motion and Emotions from Point-Light Displays in Autism Spectrum Disorders," *PLoS ONE* 7, no. 9 (September 2012): e44473, PMID 22970227, PMCID PMC3435310.

series of studies: See, for example, R. P. Hobson, "The Autistic Child's Appraisal of Expressions of Emotion," *Journal of Child Psychology and Psychiatry* 27 (1986): 321–42.

122 Research has also shown: See, for example, Michael S. Gaffrey et al., "Atypical Participation of Visual Cortex During Word Processing in Autism: An fMRI Study of Semantic Decision," *Neuropsychologia* 45, no. 8 (April 9, 2007): 1672–84; R. K. Kana et al., "Sentence Comprehension in Autism: Thinking in Pictures with Decreased Functional Connectivity," *Brain* 129, no. 9 (September 2006): 2484–93.

An fMRI study in 2008: B. Keehn et al., "Functional Brain Organization for Visual Search in ASD," *Journal of the International Neuropsychological Society* 14, no. 6 (2008): 990–1003.

"Dawson's keen viewpoint": Mottron, "Changing Perceptions."

125 I've often said: See, for example, Temple Grandin, "My Mind Is a Web Browser: How People with Autism Think," *Cerebrum* 2, no. 1 (Winter 2000): 14–22.

127 A 1981 study: Lisa D. Wiggins et al., "Brief Report: Sensory Abnormalities as Distinguishing Symptoms in Autism Spectrum Disorders in Young Children," *Journal of Autism and Developmental Disorders* 39 (2009): 1087–91.

In a 2006 study: D. L. Williams et al., "The Profile of Memory Function in Children with Autism," *Neuropsychology* 20, no. 1 (January 2006): 21–29.

The most recent: Motomi Toichi and Yoko Kamio, "Long-Term Memory and Levels-of-Processing in Autism," *Neuropsychologia* 40 (2002): 964–69.

129 Whole genome studies: Liam S. Carroll and Michael J. Owen, "Genetic Overlap Between Autism, Schizophrenia, and Bipolar Disorder," *Genome Medicine* 1 (2009): 102.1–102.7.

highly creative people: S. H. Carson, "Creativity and Psychopathology: A Shared Vulnerability Model," *Canadian Journal of Psychiatry* 56, no. 3 (March 2011): 144–53.

131 In his book: John Elder Robison, *Be Different: Adventures of a Free-Range Aspergerian* (New York: Crown, 2011).

7. Rethinking in Pictures

135 writing a paper: Temple Grandin, "My Experiences as an Autistic Child and Review of Selected Literature," *Journal of Orthomolecular Psychiatry* 13, no. 3 (1982): 144–74.

136 only a hypothesis: See, for example, Temple Grandin, "How Does Visual Thinking Work in the Mind of a Person with Autism? A Personal Account," *Philosophical Transactions of the Royal Society* 364 (2009): 1437–42.

advance copy of a book: Clara Claiborne Park, *Exiting Nirvana: A Daughter's Life with Autism* (New York: Little, Brown and Company, 2001).

141 my friend Jennifer: Jennifer McIlwee Myers interview.

The reigning champion: Jennifer Kahn, "The Extreme Sport of Origami," *Discover*, July 2006.

142 published his book: Daniel Tammet, *Born on a Blue Day: Inside the Extraordinary Mind of an Autistic Savant* (New York: Free Press, 2007).

I found an interview: Philip Bethge, "Who Needs Berlitz? British Savant Learns German in a Week," *Der Spiegel*, May 3, 2009.

143 studied the patterns: See, for example, Clifton Callender et al., "Generalized Voice-Leading Spaces," *Science* 320 (April 18, 2008): 346–48.

study classical music: Davide Castelvecchi, "The Shape of Beethoven's Ninth," *Science News* 173, no. 17 (May 24, 2008): 13.

145 physicists compared: J. L. Aragón et al., "Turbulent Luminance in Impassioned van Gogh Paintings," *Journal of Mathematical Imaging and Vision* 30, no. 3 (March 2008): 275–83.

"We expected some": http://plus.maths.org/content/troubled-minds-and-perfect-turbulence.

Jackson Pollock: Jennifer Ouellette, "Pollock's Fractals," *Discover*, November 2001.

147 FoldIt: Firas Khatib et al., "Crystal Structure of a Monomeric Retroviral Protease Solved by Protein Folding Game Players," *Nature Structural and Molecular Biology* 18 (2011): 1175–77.

149 Magnus Carlsen: D. T. Max, "The Prince's Gambit," *New Yorker*, March 21, 2011.

José Raúl Capablanca: Philip E. Ross, "The Expert Mind," *Scientific American*, August 2006.

150 "We can't help it": Michael Shermer, *The Believing Brain: From Ghosts and Gods to Politics and Conspiracies—How We Construct Beliefs and Reinforce Them as Truths* (New York: Times Books, 2011).

151 "ubiquity and appeal": Mario Livio, *The Golden Ratio: The Story of Phi, the World's Most Astonishing Number* (New York: Broadway Books, 2003).

152 a college dropout: Neal Karlinsky and Meredith Frost, "Real 'Beautiful Mind': College Dropout Became Mathematical Genius After Mugging," ABCNews.com, April 27, 2012, http://abcnews.go.com/blogs/health/2012/04/27/real-beautiful-mind-accidental-genius-draws-complex-math-formulas-photos.

a *New Scientist* article: "The Mathematics of Hallucination," *New Scientist*, February 10, 1983.

153 "People have been": http://thesciencenetwork.org/media/videos/52/Transcript.pdf.

154 a 2010 review article: Gerhard Werner, "Fractals in the Nervous System: Conceptual Implications for Theoretical Neuroscience," *Frontiers in Physiology* 1 (July 2010): 15, doi:10.3389/fphys.2010.00015.

come as no surprise: http://releases.jhu.edu/2012/10/04/jhu-cosmologists-receive-new-frontiers-award-for-work-on-origami-universe/.

the title of one paper: Maria Kozhevnikov et al., "Revising the Visualizer-Verbalizer Dimension: Evidence for Two Types of Visualizers," *Cognition and Instruction* 20, no. 1 (2002): 47–77.

155 The title of another paper: Maria Kozhevnikov et al., "Spatial versus Object Visualizers: A New Characterization of Visual Cognitive Style," *Memory and Cognition* 33, no. 4 (2005): 710–26.

Kozhevnikov said: Maria Kozhevnikov interview.

156 researchers at a neuroimaging center: Angélique Mazard et al., "A PET Meta-Analysis of Object and Spatial Mental Imagery," *European Journal of Cognitive Psychology* 16, no. 5 (2004): 673–95.

157 her original paper: Mary Hegarty and Maria Kozhevnikov, "Types of Visual-Spatial Representations and Mathematical Problem Solving," *Journal of Educational Psychology* 91, no. 4 (1999): 684–89.

she published a paper: Kozhevnikov et al., "Spatial versus Object Visualizers."

a self-report questionnaire: O. Blajenkova et al., "Object-Spatial Imagery: A New Self-Report Imagery Questionnaire," *Applied Cognitive Psychology* 20 (2006): 239–63.

an fMRI study: M. A. Motes et al., "Object-Processing Neural Efficiency Differentiates Object from Spatial Visualizers," *NeuroReport* 19, no. 17 (2008): 1727–31.

Kozhevnikov's work: See, for example, Maria Kozhevnikov et al., "Trade-Off in Object versus Spatial Visualization Abilities: Restriction in the Development of Visual-Processing Resources," *Psychonomic Bulletin and Review* 17, no. 1 (2010): 29–35.

is now widely accepted: G. Borst et al., "Understanding the Dorsal and Ventral Systems of the Human Cerebral Cortex: Beyond Dichotomies," *American Psychologist* 66, no. 7 (October 2011): 624–32.

8. From the Margins to the Mainstream

172 best-selling book: Malcolm Gladwell, *Outliers: The Story of Success* (Boston: Little, Brown and Company, 2008).

a 1993 study: K. Anders Ericsson et al., "The Role of Deliberate Practice in the Acquisition of Expert Performance," *Psychological Review* 100, no. 3 (1993): 363–406.

Consider an article: Geoffrey Colvin, "What It Takes to Be Great," *Fortune,* October 19, 2006.

175 a 2000 study: Eleanor A. Maguire et al. "Navigation-related structural change in the hippocampi of taxi drivers," *Proceedings of the National Academy of Sciences* 97, no. 3 (April 2000): 4398–4400.

177 developed a method: Sara Reardon, "Playing by Ear," *Science* 333 (September 2011): 1816–18.

180 Check out the universities: http://theweek.com/article/index/232522/virtual-princeton-a-guide-to-free-online-ivy-league-classes.

187 About fifty thousand people: Gareth Cook, "The Autism Advantage," *New York Times,* December 2, 2012.

189 see sidebar: Temple Grandin and Kate Duffy, *Developing Talents: Careers for Individuals with Asperger's Syndrome and High-Functioning Autism,* updated and expanded edition (Overland Park, KS: Autism Asperger Publishing Company, 2008).

198 an interview with Steve Jobs: Brent Schlender, "Exclusive: New Wisdom from Steve Jobs on Technology, Hollywood, and How 'Good Management Is Like the Beatles,'" *Fast Company,* May 2012.

200 Aspiritech: Carla K. Johnson, "Startup Company Succeeds at Hiring Autistic Adults," Associated Press, September 21, 2011, http://news.yahoo.com/startup-company-succeeds-hiring-autistic-adults-162558148.html.

it expanded: http://www.walgreens.com/topic/sr/distribution_centers.jsp.

201 "keep on knockin'": Savino Nuccio D'Argento interview.

202 John Fienberg: John Fienberg interview.

Acknowledgments

I wish to acknowledge all the people who made this book possible. I first want to thank my editor, Andrea Schulz, and my agent, Betsy Lerner, who helped conceptualize the structure of the book. Richard Panek, my coauthor, has been fabulous to work with. He is a superb writer who captured my voice and assembled the structure of the book. Richard's abilities in verbal and pattern thinking complemented my ability in visual thinking. We were different kinds of minds working together. His scientific knowledge was invaluable to the process. I also want to thank Tracy Roe, the copyeditor, who went beyond copyediting. She is also a medical doctor, and her input added greatly to the manuscript. Last, I wish to thank the scientists Walter Schneider, Nancy Minshew, Marlene Berhmann, and Ann Humphries at the University of Pittsburgh, Marcel Just at Carnegie Mellon, and Jason Cooperrider at the University of Utah, who did the work that made this book possible.

— Temple Grandin

In addition to the people whom Temple mentions, I would like to thank Henry Dunow, my agent, who teamed me with Temple; Virginia Hughes, whose advice on neuroimaging and genetics was invaluable; and Temple herself, an inspiring collaborator. I will miss our weekly brainstorming sessions.

— Richard Panek

Index

"acting self," 78–82, 85
AGP. *See* Autism Genome Project (AGP)
American Psychiatric Association (APA), 9, 14, 108, 109–10. *See also* Diagnostic Manual of Mental Disorders (DSM)
AMMT. *See* auditory-motor mapping training (AMMT)
amygdalae, 53
  autism and, 33, 34
  emotions and, 32, 33, 38
  TG's brain and, 31–32
Anderson, Jeffrey S., 20, 39
antidepressants
  anxiety and, 32, 38, 85–86
  link between autism and, 63–64
APA. *See* American Psychiatric Association (APA)
AQ (Autism-Spectrum Quotient) test, 104–5, 207–16
art
  object *vs.* spatial imagery and, 168–70
  pattern thinking and, 143–45
ASD. *See* autism spectrum disorder (ASD)
Asperger, Hans, 15
Asperger syndrome *(DSM* category), 15, 77, 109–11, 112
  employment and, 191–92, 200
  pattern thinking and, 142–43
Aspiritech (company), 200
associative thinking abilities, 125–28
attention-shifting problems, 90
atypical autism *(DSM* category), 111
auditory-motor mapping training (AMMT), 92–93
auditory-processing problems, 89–93
  identification of, 96–97
  TG and, 69–70
  tips for people with, 97
autism, history of, 3–20. *See also* autism spectrum disorder (ASD)
  changing psychiatric diagnoses and, 9–20
  focus on symptoms and, 11–19, 113
  phases in, 113–14
  psychoanalytic approach and, 6–9
  search for biological causes and, 6, 114
autism, neuroanatomy of. *See also* genetics of autism; neuroimaging; strengths of autistic brain
  asymmetries and, 29–32, *31*
  behavioral heterogeneity and, 37
  causal heterogeneity and, 37
  developmental abnormalities and, 27
  diagnosis and, 37–38
  directions for research on, 38–47
  research challenges and, 34–38
  size and, 30–31, 33
  structural homogeneity and, 34–36

Autism and Developmental Disabilities Monitoring (ADDM) Network, 19
Autism Center of Excellence (UCSD School of Medicine), 27
Autism Genome Project (AGP), 55–56, 58
Autism Research Centre, Cambridge, England, 104, 207. *See also* AQ (Autism-Spectrum Quotient) test
autism spectrum disorder (ASD)
  as diagnosis, 16, 104, 107–9
  impacts of *DSM-5* changes and, 111–13
  incidence of, 16–20, 34
  limitations of labels and, 101–16
"autistic behavior," 9–10
automotive exhaust exposure, 61

babbling, 45
Barnett, Jacob, 186
Baron-Cohen, Simon, 104, 207–9
Bauman, Margaret, 34
Be Different: Adventures of a Free-Range Aspergian (Robison), 131–33
Bettelheim, Bruno, 8, 67
biological motion, 121
blindness, and brain activity, 176, 177–78
bottom-up thinking abilities, 120–25.
brain anatomy, 23–27, *24*. *See also* autism, neuroanatomy of
  object *vs.* spatial imagery and, 156–57
brain trauma diagnosis. *See* high-definition fiber tracking (HDFT)
brat gene, 66
"broken-brain" studies, 26–27
Brooks, David, 203
Buffett, Warren, 172–73, 174
Burns, Mr. (genetics professor), 128–29

Capablanca, José Raúl, 149
Carlock, Mr. (teacher), 183, 194, 196
Carlsen, Magnus, 149
Carly's Voice: Breaking Through Autism (Fleischman), 80–81
Celera Genomics, 51, 55
Centers for Disease Control and Prevention, ADDM network, 19
cerebellum, and motor coordination, 27, 101, 104
cerebral cortex, 25–26, 40. *See also* visual cortex
certainty, feeling of, 124–25
CHARGE studies (UC Davis). *See* Childhood Autism Risks from Genetics and Environment (CHARGE) program
chess, and pattern thinking, 148–50
Childhood Autism Risks from Genetics and Environment (CHARGE) program, 61–62
Chopin, Frédéric, 142, 145
chunks, 150
CNVs. *See* copy number variations (CNVs)
collaboration, and three kinds of thinking, 196–97, 199
Columbia University Medical Center, 38–39, 91–92
communication deficits. *See also* language disorders; nonverbal autistic patients; speech disorders
  diagnosis of autism and, 107–9
  neuroimaging studies and, 40
computer programming, 141
connectivity
  autistic brain and, 36–37, 39–40
  development of, 45–46
  HDFT research and, 42–47, 126–27
  in TG's brain, 28, *30*, 45–46, 126–27
Cooperrider, Jason, 33, 34
copy number variations (CNVs), 56–57
corpus callosum, 103–4, 126
cosmic web, 154
Coursera, 180

Cowan, Jack, 153–54
creative thinking abilities, 128–33
crossword puzzles, 141
Crothers, Bronson, 3, 8
crying, importance of, 194
Crystal Palace, London World's Fair (1851), 125, *126*

Daly, Mark J., 58–59
D'Argento, Savino Nuccio, 200
Dawson, Michelle, 117–19, 122–23, 124, 127, 128, 169
de novo mutations, 56, 58–60
detail, attention to, 119–25
diagnosis of autism
  changing criteria for, 11–19, 53, 107–13, *vii*
  early, as important, 46–47
  Kanner and, 5–8, 9
  limitations of labels and, 101–16
  potential for biomarkers and, 37–47
  psychoanalytic approach and, 6–9
  for TG, 3–4, 9–11
Diagnostic Manual of Mental Disorders (DSM), 11–19, 53, 107–13, *vii*
  DSM-III criteria and, 14–15, 53, 107, 113
  DSM-IV criteria and, 15–16, 18, 104, 107, 109, 111–12
  DSM-5 criteria and, 19
diffusion tensor imaging (DTI), 28, 41, 42–43. *See also* high-definition fiber tracking (HDFT)
disruptive, impulse-control, and conduct disorders (*DSM* category), 110–11
DNA. *See* genetics of autism
Down syndrome, 54
DRD4-7R gene, 65–67
drugs
  cognitive responsiveness and, 85–86
  environmental triggers and, 62–64
  focus on effects and, 13
DSM. See Diagnostic Manual of Mental Disorders (DSM)
DTI. *See* diffusion tensor imaging (DTI)
dyad model, in *DSM-5*, 107–9
dyslexia, 178

Easter Seals, 201, 202
education
  accommodation of deficits and, 182–83
  exploitation of strengths and, 183–86
  special classrooms and, 182–83
  three-ways-of-thinking model and, 184–86
  useful online accessories and, 179–80
Eichler, Evan E., 58, 59
embedded-figure tests, 122
"The Emerging Biology of Autism Spectrum Disorders" (2012 *Science* article), 57
emotions
  amygdala and, 32, 33, 38
  management of, 193–94
  object *vs.* spatial imagery and, 168–69
  parental distance and, 6–9
  sensory overload and, 86–87
employment
  advice on preparation for, 187–88, 192–202
  Asperger syndrome and, 191–92, 200
  other employees and, 199–202
  pattern thinkers and, 188, 190, 198, 205–6
  picture thinkers and, 80, 190, 198, 204–5
  selling of work and, 195–96
  social impairments and, 192–96
  word-fact thinkers and, 189–90, 205–6
Encode. *See* Encyclopedia of DNA Elements (Encode)
Encyclopedia of DNA Elements (Encode), 51–52
environmental factors, 18, 60–68

error bars, 106
Exiting Nirvana: A Daughter's Life with Autism (Clara Claiborne Park), 136–39
eye contact, avoidance of, 17, 35–36, 84

faces
  bottom-up thinking and, 120–21
  cortical response to, 23, 28, 33, 35–36
feeling *vs.* behavior, 78–84. *See also* self-reporting in research
Feynman, Richard, 145, *146*
Fienberg, John, 202–3
Fleischmann, Arthur, 80
Fleischmann, Carly, 80–81, 114
flipper bridge, *164*, 165
Foldit (online game), *147*
fractals, 145, *152*–54
fractional anisotropy (FA), 41
fragile X syndrome, 55
Franklin Pierce College, 128, 129, 173–74
Freud, Sigmund, 7, 67–68
Fried, Itzhak, 32–33
frontal cortex, 25, 194
Frontiers in Neuroscience (journal), 84, 154
functional magnetic resonance imaging (fMRI), 22–23, 47–49
  biomarkers for autism and, 38–40
  sound sensitivity and, 91
  TG and, 28

Galileo, 47
Gates, Bill, 173
genetics of autism, 53–68
  AGP data and, 55–57
  directions for research in, 55
  environmental triggers and, 60–68
  fathers and, 59, 66
  junk DNA and, 50–52
  mothers and, 59, 61–62, 66
  multiple-hit hypothesis and, 58–60
  mutation studies and, 56–60
  predisposition and, 8, 60–67
  treatments for individuals and, 116
  twin studies and, 53–54
genotype, 54
Gladwell, Malcolm, 171, 173
golden ratio, *151*–52
Google, 198
grain-resolution test, 158–60
Grandin, Temple
  architectural drawings by, *12*, *138*
  associative thinking and, 125–27, 128
  bottom-up thinking and, 123–24
  brain asymmetries and, 29–32, *31*
  brain plasticity and, 178–79
  cerebellum size, 27, 101, 104
  creative thinking and, 129–30, 131
  diagnosis of autism in, 3–4, 9–11
  livestock handling designs and, *12*, 123–24, 134–35, 197
  neuroimaging studies of, 21–22, 27–33, *30*, *31*, *43*, 44–46, 126
  picture thinking and, 10–11, 134–36, 157–70
  sensory problems and, 8, 69–70, 77
  visual-spatial testing and, 157–68
grey matter. *See* cerebral cortex

hallucinations, *153*–54
handicapped mentality, 105, 192
hands-on activities, 185
HDFT. *See* high-definition fiber tracking (HDFT)
Hertz-Picciotto, Irva, 61–62
heterogeneity in autism, 37, 57–58. *See also* individual differences
high-definition fiber tracking (HDFT), 41–47, *43*, 126
"high-functioning autism." *See* Asperger syndrome
hippocampus, 53, 175
Hirsch, Joy, 34, 38, 91–92

Hobson, R. Peter, 121
Human Genome Project, 51, 55

IFOF. *See* inferior fronto-occipital fasciculus (IFOF)
ILF. *See* inferior longitudinal fasciculus (ILF)
individual differences
  focus on symptoms and, 114–16
  "label-locked thinking" and, 102–4
  sensory issues and, 73
  types of thinking and, 135
infantile autism diagnosis (Kanner's syndrome), 14
inferior fronto-occipital fasciculus (IFOF), 28, *30*
inferior longitudinal fasciculus (ILF), 28, *30*
Insel, Thomas, 34
intellectual-development disorders (*DSM* category), 111
intelligence, 6, 118–19
Internet use, 187. *See also* tablets (computer), advantages of
Irlen, Helen, 88

Jackson, Mick, 197
Jobs, Steve, 194, 198
Journal of Autism and Developmental Disorders, 109
Journal of Orthomolecular Psychiatry, 135
junk DNA, 50–52

Kanner, Leo, 5–8, 9, 67, 107
Kanner's syndrome (infantile autism), 14
Khan Academy, 179–80
Klúver, Heinrich, *153*
Kozhevnikov, Maria, 155–68

"label-locked thinking," 101–16
  being "on the spectrum" and, 101–5
  DSM definitions of autism and, 107–16
  individual differences and, 102–4
  negative effects of, 105–7, 192
  value of labels and, 107
Lancet (journal), 64
Lane, Alison, 74–76
Langdell, Tim, 121
language disorders. *See also* communication deficits; nonverbal autistic patients; speech disorders
  autism and, 17, 108
  music and, 91–92
  types of, 89–90
language-input problems, 89–90
language-output problems, 90
Lemke, Leslie, 119
Lewis, Randy, 200, 202
Livio, Mario, 152
local bias, 122
"The Long-Range Interaction Landscape of Gene Promoters" (*Nature* article), 52
long-term memory, 127–28
Lord, Catherine, 19

McKean, Thomas, 88
magnetic resonance imaging (MRI), 22, 29–33. *See also* diffusion tensor imaging (DTI); functional magnetic resonance imaging (fMRI); neuroimaging
Maguire, Eleanor, 175
manners, 194–95
Massachusetts Eye and Ear Infirmary, 177–78
maternal obesity, 62
mathematics
  algebra *vs.* geometry thinking in, 147–48, 183–84
  pattern thinking and, 140, 142, 143–48, 151–53, 186–87
"meaning blindness," 90
Meares, Olive, 88n

Medical Investigation of Neurodevelopmental Disorders (MIND) Institute, University of California, Davis, 61–62
medical treatment, and labels, 105–6
medication. *See* drugs
memory. *See also* short-term memory; visual memory
  associative thinking and, 127–28
  brain structures and, 29, 32–33
Mendel, Gregor, 128
mental retardation, and autism, 16–17
mentors, value of, 196
Miller, Sara R. S., 141
mind, theory of, 72, 86–87
mismatch field (MMF) studies, 91
mitochondrial disease, 64–65
MMF studies. *See* mismatch field (MMF) studies
Monger, Christopher, 197
Mottron, Laurent, 117, 124, 127
movement sensitivity, 75, 76. *See also* biological motion
music, 143, 145, 186
music therapy, 91–93
Myers, Jennifer McIlwee, 141

Nature (journal), 52, 58–60
Nelson, Stanley, 57
neurexin protein, 55, 58
neuroimaging. *See also* autism, neuroanatomy of
  interpretation and, 48–49
  limitations of, 23, 47–49
  object *vs.* spatial imagery and, 156–57
  research on autistic brain and, 34–47
  TG's brain and, 21–22, 27–33, *30*
  types of, 22–23
  visual *vs.* spatial imagery and, 156, 157
neuroligin protein, 55, 58
Neuroscience and Biobehavioral Reviews (journal), 84
new experiences, value of, 187–88
New Scientist (journal), 128–29, 152–53
New York Times, 50–52, 203
nonverbal autistic patients
  diagnostic criteria and, 108
  intelligence testing and, 118
  music and, 91–93
  responsiveness and, 77–80, 91–92
  self-reports and, 77–82
  sensory disorders and, 71
  strengths and, 119
  technology and, 77, 78–80

object visualizer, 154–55. *See also* visual (picture) thinking
object *vs.* spatial imagery, 154–57, 168–70
obsessions, 188–89
occipital cortex, 25. *See also* visual cortex
O'Hare Airport United terminal, 125, *126*
Ohio State University, 74–76
olfactory sensitivity, 94–95, 98
one-on-one engagement, 4
origami, 140–42, 154
overconnectivity theory, 36
overresponsiveness to sensation, 74, 83

Padgett, Jason, *152*
parenting. *See also* genetics of autism
  identification of child's strengths and, 181
  negative environments and, 6–9, 66–67
  preparation for employment and, 187–88
  TG's mother and, 3–4, 8–9
parietal cortex, 25, 29, *31*
Park, Jessica ("Jessy"), 136–39, *138*, 155
patternicity, 150
pattern thinking. *See also* picture thinking; verbal (word-fact) thinking; visual (picture) thinking

as category, 135–40, 150–52
chess and, 148–50
education and, 186–87
employment and, 188, 190, 198, 205–6
examples of, 141–50, 152–54
mathematics and, 140, 142, 143–48, 151–53
object *vs.* spatial imagery and, 154–57, 168–70
origami and, 140–42
research and, 154–57
PDD. *See* pervasive developmental disorders (PDD)
pervasive developmental disorders (PDD), 14–15, 109. *See also* Asperger syndrome; autism spectrum disorder (ASD)
pervasive developmental disorders not otherwise specified (PDD-NOS), 15, 17–18, 109, 111, 112
pesticide exposure, 62
pharmacological treatment. *See* antidepressants; drugs
phenotype, 54
Physical Therapy (journal), 76
picture thinking. *See also* pattern thinking; verbal (word-fact) thinking; visual (picture) thinking
education and, 184–85
employment and, 80, 190, 198, 204–5
plasticity of brain, 175–82
Pollock, Jackson, 145
prefrontal cortex, 25
psychoanalytic approach to diagnosis, 6–9
public views of autism, 17, 199–203
Pythagorean theorem, 147–*48,* 152

Rajarshi Mukhopadhyay, Tito, 78–80, 81–82, 119, 120
Raven's Progressive Matrices, 118–19, 139–40
reading. *See* visual-processing problems
"Recognition of Faces: An Approach to the Study of Autism" (1978 study), 120–21
"refrigerator mother" concept, 6–8
research
algorithmic approach and, 39–40
on brain plasticity, 177–78
DSM-5 diagnostic criteria and, 112
focus on symptoms and, 114–15
HDFT and, 41–47
label-locked thinking and, 106, 114–16
on neuroanatomical biomarkers, 37–47
on sensory problems, 71–73, 74–76, 106
technology and, 47, 48, 77–80
young subjects and, 40–41
responsiveness, and sensory problems, 83–87
responsiveness gene, 67. *See also* environmental factors
Rett syndrome, 54–55
risperidone (Risperdal), 86
Robison, John Elder, 131–33
Rosenhan, David, 13
rudeness, 7, 193, 195

savants, 119, 141–42
Schaefer, G. Bradley, 57–58, 65
schizophrenia, 9, 11, 13–14
Schneider, Walter, 42–47, 115, 127
Science (journal), 50, 56, 57
self-reports. *See also* Grandin, Temple
elicitation problems and, 77
focus on symptoms and, 114–15
sensory problems and, 76–84
tablets and, 77–78
types of visual thinkers and, 157
sensory interaction, study of, 123–24

sensory problems in autism
attitudes toward, 73, 106–7
identification of, 95–98
impacts of, 71, 72–73
research on, 71–73, 74–76, 106
responsiveness and, 83–87
self-reports on, 78–80, 81–82
sensory domains and, 75, 76, 87–98
subtypes of, 74–76
TG and, 8, 32, 69–70, 77, 83
tips for people with, 96–98
"Sensory Processing Subtypes in Autism: Association with Adaptive Disorders" (Lane, et al), 74–76
sensory seeking behaviors, 74, 75
Šestan, Nenad, 116
SHANK2 gene, 59–60
SHANK3 protein, 55, 58
Shermer, Michael, 150
Short Sensory Profile (research tool), 75–76
short-term memory, 126, 150
TG and, 10, *31*, 126–27, 161
Silicon Valley employees, 105, 182
singing, therapeutic effects of, 91–93
smells, sensitivity to. *See* olfactory sensitivity
social communication disorder (*DSM* category), 109–10
social impairments
antidepressants and, 86
diagnosis of autism and, 107–10
employment and, 192–96
rudeness and, 7, 193, 195
sensory problems and, 72–73, 86–87
training and, 103, 190–91, 192–93, 194–95
The Social Network (movie), 203
Soulières, Isabelle, 119
sound sensitivity. *See* auditory-processing problems; sensory problems in autism
spatial-relations test, 160–63, *162*, 167–68
spatial short-term memory, 127
spatial visualizing, 154–55. *See also* pattern thinking
spatial *vs.* object imagery, 154–57, 168–70
Specialisterne (company), 200
speech disorders, 17, 38, 109
TG and, 3–4, 45–46
Squidoo (website), 173, 174
Stahl, Lesley, 46–47
Starry Night (van Gogh painting), *144*, 145
State, Matthew W., 58–59, 116
strengths of autistic brain, 117–33, 135
associative thinking ability and, 125–28
bottom-up thinking ability and, 120–25
brain plasticity and, 175–82
collaboration and, 196–99
creative thinking ability and, 128–33
identification of, 181–82
intelligence and, 118–19
memory and, 127–28
"Strong Association of De Novo Copy Number Mutations with Autism" (2007 *Science* article), 56
structural MRI. *See* magnetic resonance imaging (MRI)
sudoku, 141
symptoms, focus on, 11–19, 113, 114

tablets (computer), advantages of, 77–78, 179
tactile sensitivity. *See* touch sensitivity
Tammet, Daniel, 85, 142, 155
taste sensitivity, 94–95, 98
Taylor, Richard, 145

technology
nonverbal autistic patients and, 77, 78–80
research and, 47, 48, 77–78
Temple Grandin (HBO movie), 197
temporal cortex, 25–26
temporoparietal junction (TPJ), 35
10,000-hour rule, 171–74
thinking, types of. *See* pattern thinking; strengths of autistic brain; three-kinds-of-minds approach; verbal (word-fact) thinking; visual (picture) thinking
Thinking in Pictures (Grandin), 134–35, 137, *viii*
Thinking Outside the Brick exercise, 130–31
"thinking self," 78–82, 85
3-D drawing tools, 180
3-D printers, 180
three-kinds-of-minds approach, 170. *See also* pattern thinking; picture thinking; verbal (word-fact) thinking; visual (picture) thinking
education and, 182–87
employment and, 187–202
touch sensitivity, 93–94, 97–98
TPJ. *See* temporoparietal junction (TPJ)
triad model, in *DSM-IV,* 107–9
Triplett, Donald, 5, 190
tuberous sclerosis, 55
twin studies, 53–54

underconnectivity theory, 36–37
underresponsiveness to sensation, 74, 83
Université de Caen and Université René-Descartes, France, 156
universities, free courses from, 180
University of Amsterdam, 36–37
University of California, San Diego, 27
University of Louisville, 40
University of North Carolina at Chapel Hill infant study, 41
University of Pittsburgh, 28, *30,* 41–42, 126
University of Utah, 29, *31,* 39
University of Washington, 58

vaccination, and autism, 64–65
van Dalen, J. G. T., 85
van Gogh, Vincent, *144,* 145
Venter, Craig, 51
verbal (word-fact) thinking. *See also* pattern thinking; picture thinking; visual (picture) thinking
autism and, 136, 141
as category, 135, 154
education and, 180, 185–87
jobs involving, 189–90, 205
spatial testing and, 155–56, 167
TG and, 11, 136
video games, 188–89
vision, restoration of, 176
visual cortex
aural cues and, 91
blind subjects and, 177–78
brain anatomy and, 25
hallucinations and, 153–54
injury to, 27, 152
plasticity and, 176, 177–78
TG and, 28, 44
visual memory, 28, *30,* 36
visual-processing problems, 87–89, 134
brain plasticity and, 178
identification of, 95–96
tips for people with, 96
visual-spatial tests, 157–68
visual (picture) thinking
pattern thinking as category and, 135–41, 150–54

visual (picture) thinking (*cont.*)
spatial *vs.* object imagery and, 154–57, 168–70
TG and, 10–11, 134–36, 157–70
vitamin supplementation, 61
VVIQ (Vividness of Visual Imagery Quotient) test, 157–58

Walgreens, 200
Wan, Catherine Y., 92–93
"weak central coherence," 120
Wechsler Intelligence Scale for Children, 118–19
white matter, 26, 29, 30, 41
Williams, Donna, 85, 88, 90, 114, 121
Wiltshire, Stephen, 119
Wing, Lorna, 15
word-fact thinking. *See also* pattern thinking; picture thinking; verbal (word-fact) thinking; visual (picture) thinking
education and, 180, 187
employment and, 189–90, 205
"World Changing Too Fast" (2009 research paper), 84–85

Yale University School of Medicine Child Study Center, 58

Zuckerberg, Mark, 203